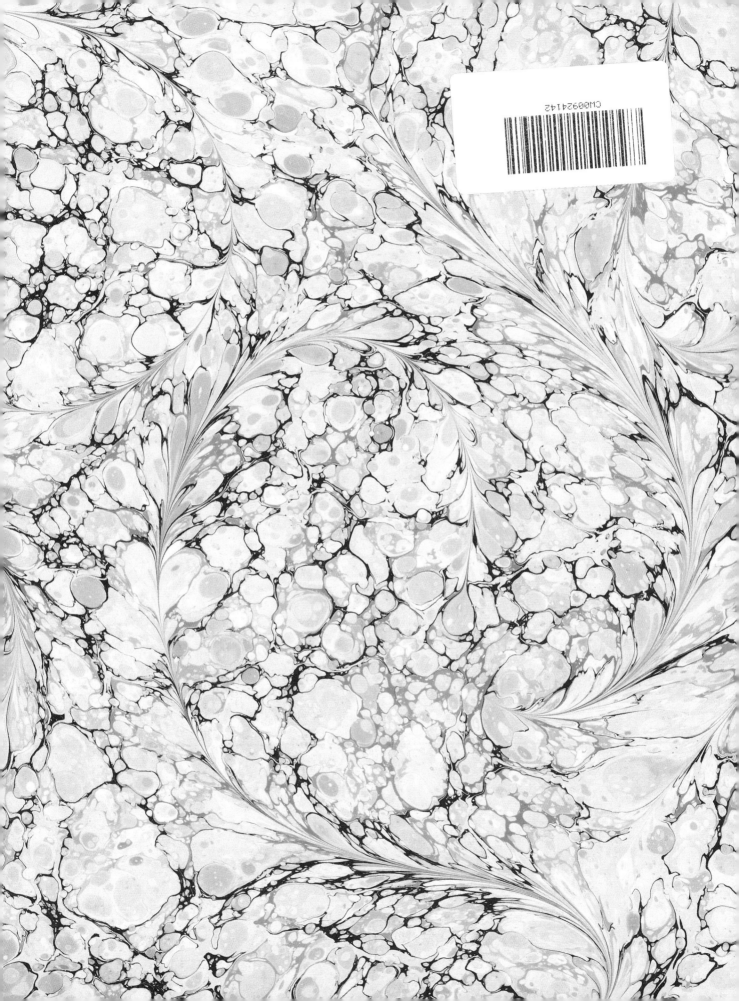

BMW MOTORCYCLES IN WORLD WAR II

Janusz Piekalkiewicz

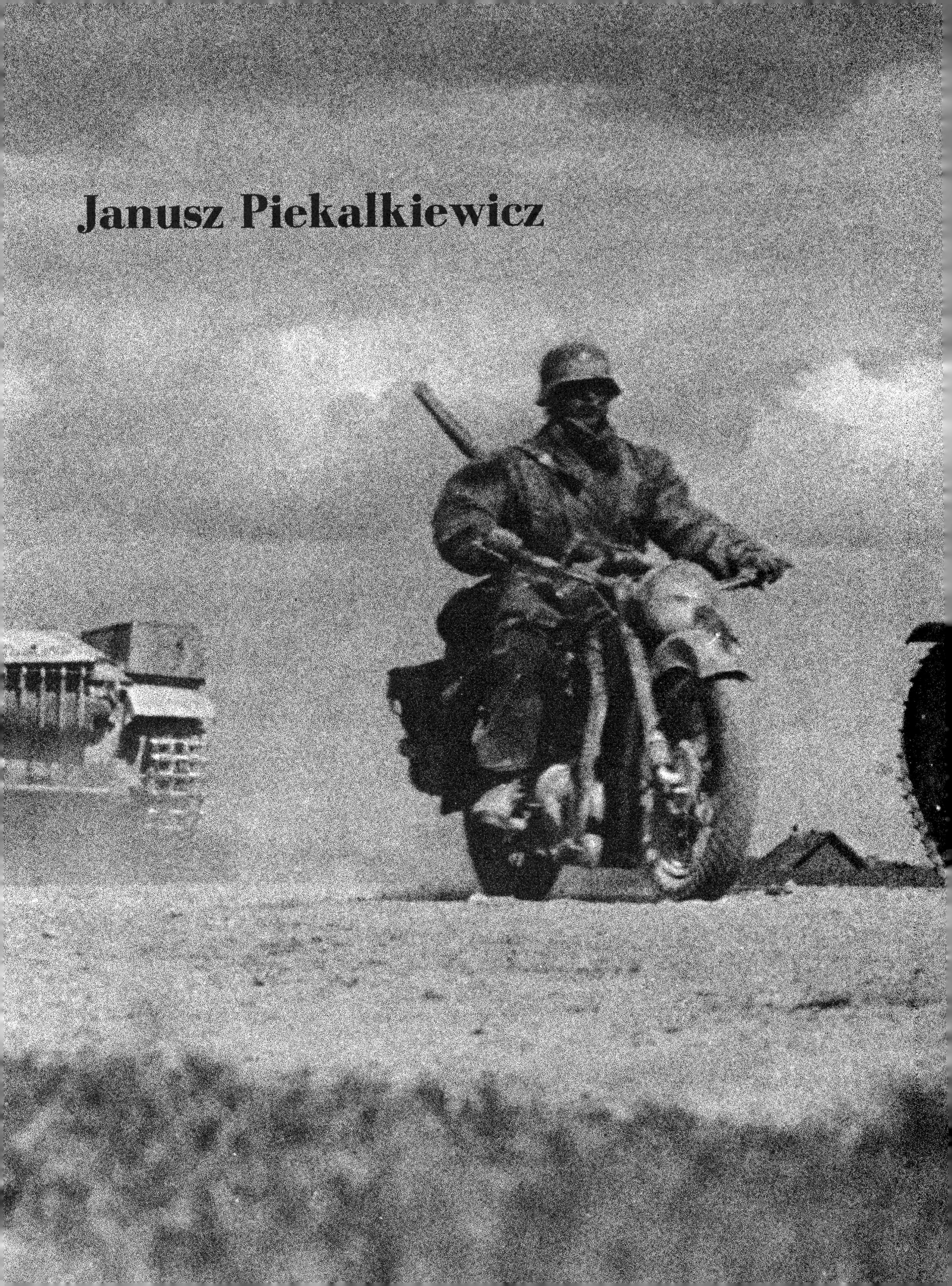

BMW MOTORCYCLES
IN WORLD WAR II
R12/R75

Schiffer Military History
Atglen, PA

Cover Drawing: Carlo Demand.

Translated from the German by Dr. Edward Force.
Central Connecticut State University.

Copyright © 1991 by Schiffer Publishing, Ltd.
Library of Congress Catalog Number: 91-60854

All rights reserved. No part of this work may be reproduced or used in any forms or by any means – graphic, electronic or mechanical, including photocopying or information storage and retrieval systems – without written permission from the copyright holder.

Printed in China
ISBN: 978-0-88740-306-4

This book was originally published under the title,
Die BMW Kräder R12/R75 im Zweiten Weltkrieg,
by Motorbuch Verlag, Stuttgart.

We are interested in hearing from authors with book ideas on related topics.

Published by Schiffer Publishing Ltd.
4880 Lower Valley Road
Atglen, PA 19310
Phone: (610) 593-1777
FAX: (610) 593-2002
E-mail: Schifferbk@aol.com
Visit our website at: www.schifferbooks.com
Please write for a free catalog.
This book may be purchased from the publisher.
Try your bookstore first.

CONTENTS

Foreword	7
The BMW R 12/R 75 Motorcycles in World War II	9
Technical Data	159
Bibliography	189
Acknowledgements	189
Photo Credits	189

FOREWORD

Lauded when it appeared during the war as "the off-road motorcycle of the future", the BMW R 75 was not just a powerful "workhorse" in terms of appearance. Its reliability and easy handling gave it something of the untiring patience of the horse in whose hoofprints it followed. The riflemen as well as the messengers who changed from an earlier "Krad" to the R 75 acquired a true comrade-in-arms in this machine. And this R 75 with its sidecar drive was a great advance in wartime motorcycle production by the Bayerische Motoren Werke AG.

Whether crossing water or swampy fields, rolling along in a column by the hour, conquering steep slopes or moving fast along the Autobahn, this heavy machine handled it all without sacrificing its safe handling. It provided service with the same stamina in the sands of North Africa as on the snow-covered plains of Russia.

It was this heavy machine that made possible the motorcycle rifle battalions of the German *Afrika-Korps* to make fast forays in the trackless wastes of Libya or Cyrenaica. Thus it was the motorcycle riflemen of the *Knabe* Advance Unit that were among the first German soldiers who advanced into the Land of the Pharaohs.

On the Russian front as well, many a surprising success was attributed to the R 75 units, often acting alone. Often they were the first — far ahead of the armored scout vehicles — to appear suddenly in a town or at a river crossing and create confusion.

The veterans who took part in the campaigns of the German Wehrmacht in World War II remembered the motorcyclists vividly. It was a special breed of men who — in heat and dust, frost, snow or mud, exposed almost without protection — nevertheless retained their humor and their good spirits. Above all, they were known as helpful comrades and understanding souls. After all, the capability of a cycle messenger is important to the survival of every unit. What good are the best orders if they don't arrive — if telephone lines are destroyed and radio silence is ordered?

One must realize that even the system of higher command posts, the brain of the division, had to move its position often in mobile warfare.

To this day, in addition to the memories of those days, the machines and the men who drove them, a number of heavy BMW motorcycles have been preserved, finding enthusiastic collectors in East and West alike.

The photos in this volume tell us not only of the hard times and sorrows, but also of many a happy hour for these men whose fate it was to spend their best years in the saddle of a BMW R 75.

Janusz Piekalkiewicz

1

"It was on the Don, and every trace of a road had turned to mud in a cloudburst", reported *Obergefreite* Schmid, motorcycle messenger of the 4th Panzer Army. "No vehicle could move forward even a few meters in this "shaving soap", but the message had to get through. I put snow chains on the two drive wheels, shifted to the lowest gear and let the motor run slowly, and got through. Nobody believed I could make it with my BMW." He was driving the heavy "Krad" with sidecar drive "for swamp, sand and snow", the 750 cc R 75 — made by the Bayerische Motoren Werke AG.

A long difficult path led to the initial series production of this model — this martial vehicle that could handle even the roughest action.

As early as 1934 a BMW motorcycle and sidecar with drive to both rear wheels took part in a winter test run and attracted widespread attention in the trade. At first, though, this idea was not put into production, since its development was not regarded as having been completed. But foreign motorcycle manufacturers, inspired by the first experiences gained by BMW, decided to build military cycles with sidecar drive, but at that time none of the designs of that type proved to be ready for production. One of the first decisive steps in this area had been taken years before by a man who had very little to do with motorcycles until then: Max Friz, the aircraft-engine builder.

To him we also owe the basic conception of the four-stroke motor as an opposed two-cylinder unit with rear-wheel belt drive. Friz, who created the first practical German motor for sport planes in 1912, developed high-performance motors for the fighter planes of the Imperial German air forces in World War I. With the last motor he designed during the war, the "BMW IV High-Altitude Motor", what was then a remarkable altitude of 9760 meters was reached on June 17, 1919.

When the victorious powers forbade the further construction of airplanes in Germany, Max Friz developed motors for boats. In 1922 Dr.-Ing. M.C. Friz, later the chief designer and director of the Bavarian Motor Works, developed for BMW the M2 B 15 motor, the first air-cooled opposed motor built to power stationary machinery. This motor made by Friz was then used by the Victoria Works in Nürnberg for a different purpose: as a powerplant for their motorcycles.

The Bayerische Flugzeugwerke AG, which was banned from building aircraft motors, gradually got the idea of equipping their motorcycles with the Friz-BMW motor too. Thus the cycle with the impressive name of "Helios" was born. It was still built after the merger of the Bayerische Flugzeugwerke AG with the Bayerische Motoren Werke AG.

The later two-cylinder BMW cycles with their horizontally opposed motors and universal-joint drive are the high point of Dr. Friz's creative achievement. Until the end of World War II, he had a large share in the development of all types of BMW vehicles. In 1932 the Bayerische Motoren Werke AG, which had devoted itself to building aircraft and stationary motors until then, put its first motorcycle on the market. It was the BMW R 32, with a 500 cc motor producing 8.5 HP at 3300 rpm. It bore the blue-and-white BMW emblem and attracted attention with its modern design. Its progressive technical features stood out from the ranks of German and foreign motorcycles. Especially striking in the R-32 cycle for example, was the opposed motor, mounted transversely, with universal-joint drive to the rear wheel. BMW Director Max Friz, the model's creator, built the foundation for all subsequent BMW models with this design.

Successful competition in various sporting events of 1924 and winning the German Road Championship are the clearest proof of the superior performance of the new BMW cycles. To be able to continue this series of victories, the firm brought out the R 37, an overhead-cam 500 cc sport machine with 16 HP at 4000 rpm, in 1925, as well as a 250 cc, 6.5 HP, one-cylinder cycle, the R 39. The R 42, R 47, R 52 and R 57 models, all 500 cc motorcycles, continued the basic design features for several more years.

The 16 HP BMW R 37 was the sport cycle of the year in 1925. The front-wheel suspension by a short leaf spring, reminiscent of automobile construction, and the easily used tank switch stand out among its features. Not many of these models were built, however they did score noteworthy victories.

1928 brought a displacement change from 500 to 750 cc in the R 62 model, which produced 18 HP at 3400 rpm, and the overhead-cam R 63 with 24 HP at 4000 rpm. In 1929 BMW decided to give its motorcycles a more rigid chassis and built the R 11 — with 18 HP at 3400 rpm, and the R 16 with 26 HP at 4000 rpm and a pressed-steel frame. In 1928 BMW was the only German competitor in the British international six-day race, finishing without losing a single penalty point. The Eilenriede race, the Targa Florio, the Austrian Tourist Trophy, the Grand Prix of Europe and numerous other races were all won by BMW at that time, as well as the German Road Championship.

One of the outstanding achievements of the BMW motorcycles was the record run made by Ernst Henne in the autumn of 1929, in which he won Germany's first absolute world speed record for motorcycles. He used the 500 and 750 cc BMW supercharged cycles with aerodynamic bodywork, and reached a top speed of 216.75 kph.

Ernst Henne was able to set new absolute world speed records seven times, and on his last run in 1937 he reached a top speed of 279.503 kph with the streamlined 500 cc machine, which had been developed by Director R. Schleicher to produce 100 HP. The BMW R 2 cycle of 1931 was equipped with a pressed steel frame and fork and an overhead-cam one-cylinder 200 cc motor that produced 6 HP. This model met the demand for a lower-priced, more economical machine. Shock absorbers on its front suspension, an aluminum cover to keep the valves dust-free, a ball gearshift and a nicely formed saddle tank gave this machine a special aura.

In 1933 J. Stelzer set out on winter testing in the foothills of the Bavarian Alps on a BMW cycle that was equipped for the first time with a sidecar. A year later, Ernst Henne also used this two-wheel drive system for his record run in Hungary. In both cases BMW was completely successful.

A significant change in appearance, along with a considerable improvement and stabilization of handing characteristics, characterized the BMW machines of 1935 when the front-wheel leaf springs were replaced by telescopic springs on the fork.

In exhaustive testing, a suspension system was developed in which the wheel did not swing but moved straight up and down, which made it possible to build the suspension system as small as possible and house it, with hydraulic shock absorbers, in a very sleek front fork. The BMW R 12 model, with 18 HP at 3400 rpm, and the R 17, with 33 HP at 4500 rpm, were the first to be equipped with this new system, which has been retained with only minor changes to this day. The R 12 was produced in greater numbers than ever before and was one of the most reliable Wehrmacht cycles.

In 1936 the tube frame, already used in the first BMW motorcycles, was revived for the sleek-lined R 5 model, with 2400 HP at 5800 rpm, and the R 6, with 18 HP at 4800 rpm. The one-cylinder machines also made technical progress; they were the 200 cc R 20 and the 350 cc R 35, with the reliable telescopic fork and tube frame.

But it was the last prewar models, the R 51, R 61, R 71 and R 66 of 1938, that achieved the best performance; they featured additional rear suspension, a swinging seat and a fully new, unique type of design. These production models show the highest level of BMW motorcycle technology before World War II.

June 16, 1939 was a great day in BMW history. European champion Georg Meier, on a BMW machine, was able for the first time to win a decisive victory over Britain's best riders in the 500 cc senior class of the Tourist Trophy. When the German military leadership urged the development of a heavy cross-country motorcycle in 1939, BMW was ready to make practical suggestions.

In the spring of 1941 the new BMW R 75 heavy motorcycle, producing 26 HP at 4900 rpm, was ready. It was a new multipurpose cycle whose most important features were sidecar wheel drive, a locking differential, cross-country and reverse gears, a felt air filter, a new chassis, off-road tires and excellent handling. The motor of this new BMW R 75 was a transverse overhead-cam two-cylinder type displacing 750 cc and attaining 26 HP; higher top-end performance was avoided in favor of better low-end torque to make for a less sensitive motor. The combustion chambers were hemispherical. The camshaft, generator and magneto were driven by aluminum gears which were meant to fit the smooth engine block and provide noise-free operation. A voltage regulator controlled the independent ignition adjustment at changing engine speeds.

A Graetzin carburetor was attached directly to each cylinder, so as to avoid long fuel lines. The air for both carburetors was cleaned by a single moist-air filter with a pre-filter, an oil strainer and oil sump, by which even the smallest particles of dust could be removed. The transmission had a four-speed gearbox with a dog clutch and crown wheels constantly in contact; it ran very quietly and experienced very little wear. By using a shift lever as a pre-selector, the gear ratios could be made smaller in order to give the machine greater low-end power in rough country. The reverse gear, a new feature of this transmission, provided hitherto unknown off-road mobility and allowed a quick shift into reverse when stuck in mud or snow.

Power transmission to the rear wheel was done, as on all BMW motorcycles, by a driveshaft held by a shock-absorbing rubber cross link on the gearbox.

The heavy BMW R 75 built in 1941 was equipped with a power-dividing crown-wheel differential attached to the rear wheel drive to equalize the varying speeds of the two driven wheels resulting from different road conditions. Thus the steering of this rig was as easy as that of a four-wheel vehicle and avoided premature tire wear.

If the cycle was stuck in mud, sand or snow and one wheel was spinning freely, the equalizing gear locked and both drive wheels ran on fixed axles, so that the machine could overcome such hindrances.

Another change was made to the telescopic front fork in terms of its outer covering. The two telescopic fork covers enclosed the suspension mechanism so as to keep out most dust, but the smallest dust particles were not kept out. In the new 1941 model, the fixed part of the fork was attached to the movable lower part by a rubber sleeve so that no air or dust could get in. The rubber sleeve was added to all the models that did not already have it. Thus equipped, the 1941 BMW R 75 with its driven sidecar wheel was the cycle for all terrains and all climates. The front fork of the R 75 was the strengthened telescopic spring fork of the previous BMW models, with double-action hydraulic shock absorbers. The chassis consisted of an enclosed lattice-girder construction which could be dismantled into several individual parts and thus was easy to repair, though several triangular braces and a central box section gave it considerable rigidity.

The interchangeable drive wheels had stub axles for easy changing. They were strong spoked wheels with deep-bed rims and cross spokes. The tire size of 4.50 x 16, unusual for a drive wheel but the same size as used on the VW *Kübelwagen*, afforded better off-road traction with little wear, increased comfort and lengthened tire life. The rear wheels and sidecar wheel had hydraulic brakes, with safe and equal braking effect on both drive wheels.

The fuel tank with its 24-liter capacity was sufficient for a range of 380 kilometers at an average consumption of some six liters per 100 kilometers.

The sidecar hull was mounted on a rectangular frame with leaf springs and rubber cushions. The major shocks of rough country, though, were absorbed by a sidecar wheel mounted on a swinging suspension arm. The vertical movement of the whole sidecar activated an attached tube spring system; inside the transverse tube was a torsion bar that transmitted the power from the differential to the gears in the swinging arm.

The cumulative effect of the many special features allowed high performance in rough country that had previously not been passable by motorcycle. High clumps of grass, deep ruts made by trucks, and loose stone were no longer a hindrance, thanks to the high ground clearance of 275 mm under the sidecar.

Of even greater importance than the sustained speed of 95 kph was the marching speed of 3 kph that enabled these machines to accompany columns of marching men on long stretches without overheating the motor.

Polish campaign, September 6, 1939: The motorcycle riflemen of the I. Inf.Reg. 151 of the 61st Infantry Division in combat in the streets of the town of Pultusk.

Polish campaign, September 5, 1939: The motorcycle messengers and riflemen of an armored unit at rest. In the small Galician village of Nizankowice, near Przemysl, which the spearhead has reached, the battle goes on. In the background is a former Czech 38 (t) tank, Type TNHP.

Polish campaign, September 6, 1939: The market place of Pultusk is won. In the foreground are two motorcycle riflemen of a Luftwaffe unit with a heavy 30 HP, 600 cc OHV 1938 BMW R 66 unit (right) and a 22 HP, 750 cc SV 1938 BMW R 71 (left), the latter with a Felber civilian sidecar and civilian paint.

Polish campaign, September 8, 1939: A motorcycle rifleman of an SS front-line unit on a heavy BMW R 12 1935-41, 20 HP, 750 cc SV motor, in the hard-hit town of Pinczow.

Polish campaign, September 9, 1939: a Luftwaffe *Obergefreiter* and members of a police battalion with a local man on a heavy BMW R 12 1935-41, 20 HP, 750 cc SV motorcycle.

Before Warsaw, September 9, 1939: The fate of a motorcycle rifleman in the advance unit. The medics remove the fatally injured passenger.

Polish campaign, September 1939: An unusual vehicle, The heavy BMW R 12 1935-41, 20 HP, 750 cc SV, as towing tractor and personnel carrier for a light infantry gun while changing position.

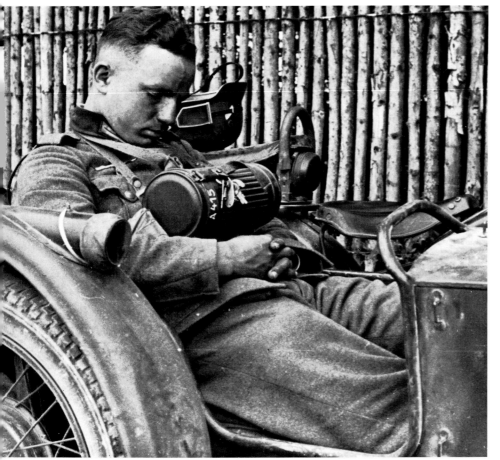

Polish campaign, September 1939: A well-earned rest between two missions in the sidecar of a BMW cycle.

Germany, winter 1940: Motorcycle rifle unit drivers dismount during field training. At right are the heavy BMW R 12 1935-41, 20 HP, 750 cc SV cycles.

Germany, winter 1940: A motorcycle rifle unit begins a marching drill on the heavy BMW R 12 1935-41, 20 HP, 750 cc SV machines.

A motorcycle repair shop in Germany, 1940: A mechanic works on the right heating duct of the carburetor warming system of a heavy BMW R 12 1935-41, 20 HP, 750 cc SV machine, which still bears its Berlin (1A) registration number.

The front end of a heavy BMW R 12, 1935-41, 20 HP, 750 cc SV cycle of an intelligence company. A mechanic overhauls the motor.

Germany, North Sea coast, spring 1940: In the sidecar of the heavy BMW R 12 1935-41, 20 HP, 750 cc SV cycle sits a war correspondent from the Luftwaffe.

Germany, North Sea coast, spring 1940: The civilian technicians of a Luftwaffe test center with a heavy BMW R 12 1935-41, 20 HP, 750 cc SV cycle.

Germany, spring 1940: "Mount!" for cross-country combat training with heavy BMW cycles.

2

In the spring of 1941 the new BMW R 75 model began to appear on the dusty desert roads of Libya. The heavy machine, which was originally also supposed to serve *Fallschirmjäger* units as a "draft horse" to pull their recoilless light guns, was soon released from this service. At the first test run it was seen that the idea was not so good in practice: the weight of the gun trailer lifted the cycle's front wheel off the ground and the rig could not be steered.

The mighty machine received its baptism of fire with the German *Afrika-Korps* and soon won the almost legendary fame that it kept well beyond the war's end. It became every motorcycle rifleman's and messenger's favorite piece of equipment. And these were the very men who knew how to treasure its advantages — and on which their lives often depended.

In the spring of 1941 the units of the German *Afrika-Korps* were in Libya and Cyrenaica, faced a hard fighting enemy. During the fighting in the stony and sandy wastes of North Africa, the heavy BMW R 75 machines proved themselves best of all, and the fast cycle troops contributed to Rommel's success. Even one of the men who knew the desert best among the British Army in that area, Lieutenant General Sir Richard O'Connor, began to become aware of that.

On April 2, 1941 he was ordered to the Cyrenaican front along with Lieutenant General P. Neame to straighten out a miserable situation. Four days later, on the evening of April 6, the two desert warriors drove to their headquarters at Tmimi. Their Humber staff car got lost in the trackless terrain before they finally found the road to Derna.

Suddenly a German motorcycle rifle patrol of the *Ponath* Group appeared. In a short fight, the driver of the Humber lost his life and the two lieutenant generals and the commander of the 11th Hussars, Brigadier Combe, were captured by the motorcycle men. Lieutenant General O'Connor was still wearing the Italian Medal of Merit, which he had received in 1918 after the battle of the Piave.

A fourth English general, Gambier-Parry, was taken prisoner near Mechili the next day, and Brigadier General Remington surrendered in Derna.

Rommel, whose heavy artillery was still on the quay at Naples, and who knew that he would have no easy time taking Tobruk, gave two of his fast battle groups the command to push farther eastward — beyond Tobruk. It was Motorcycle Rifle Battalion 15 under *Oberstleutnant* Gustav-Georg Knabe, supported by the Panzerjäger Unit 33 with an 88mm flak battery. The second unit was Reconnaissance Unit 3 of Baron von Wechmar.

The three rifle companies of Motorcycle

Rifle Battalion 15 with their 198 heavy cycles, 75 cars and trucks and 630 men met with the venerable transport "Alicante" in Tripoli on April 5. Instead of their promised week of rest, they received Rommel's order here: "Move out at once in the direction of Benghazi."

The little battle group worked its way laboriously through the loose sand of the dusty road. On each of the heavy cycles was a three-man crew with an MG 34 machine gun, ammunition cases, three packs, canisters of fuel and water, and three rifles. And all of that at a temperature of 50 degrees Celsius. The column rolled ahead according to orders: Stop every ten kilometers and let the motors cool against the wind. At every stop, thousands of flies attacked them; while underway it was dust, stones and camel thorn. They were some 50 kilometers west of Capuzzo when night suddenly fell. Patrols with radios were sent forward. Six years later they came back by way of Canada.

At sunrise on Easter morning, the cry of "Happy Easter!" from some comrades awakened the riflemen. But instead of their scouts, it was the British. Three cycles and their crews were destroyed. Only the 88mm flak gun restored order.

The column hurried on to the east. Soon the silhouette of the desert fort "Capuzzo" shimmered in the sun. In the ruins of this once-roomy Italian fortress with its officers' and men's quarters they found stores of British rations. With baked ham, pineapples from the refrigerator and whisky they celebrated Easter in a big way.

The report of the *Knabe* advance group went on to say: "At 12:10 Capuzzo, other than Tobruk the last Italian fort in North Africa still in British hands, which formed a dominant, strongly fortified support point in the completely unbroken desert landscape, was taken, with only one 88mm flak battery and without our own armored forces, from an enemy heavily armed with tanks and armored scout cars as well as mobile artillery. Flak and Panzerjäger units had to substitute for the lacking artillery as best they could.

Our success is due largely to their determined action and good results. To prevent a premature positioning of the hastily withdrawing enemy, the commander of the advance group ordered an immediate move forward to Sollum.

"The 3./K.B. 15, followed on trucks as a reserve until then, was moved forward against Sollum, reinforced by Panzerjäger Unit 33 and two 20mm flak guns of the 4th flak battery. An armored reconnaissance troop, made available to the advance unit by A.A. 3, with three armored scout cars and an armored radio vehicle, set out for Sollum under the command of *Oberleutnant* Moder (Staff Pz.Jg. 33) and reported border posts and road passage free of enemies, Sollum and the heights to the south occupied by the enemy.

"At 1:25 P.M. the *Knabe* advance unit became the first German troop to cross the Egyptian border. Through the encircling attack on Sollum, a scattering of the enemy attack fire was attained. South of the Capuzzo-Sollum road the 2nd Company of KB 15, already used at Capuzzo to encircle from the right, and now exposed to several enemy attempts to launch tank attacks, moved farther to the east in the same direction and eliminated the threat of a flank attack from the south, while along the coast to the north, a motorcycle rifle column of the 1st Company of KB 15, reinforced by a Panzerjäger column, was sent against Sollum. The 88mm flak battery watched over the proceedings and supported the attack from an open firing position northwest of Sollum. After a short battle, the enemy was forced to evacuate Sollum, the town itself was mopped up and the heights on the coast were occupied. At 3:45 P.M. the leader of the advance unit reported to the D.A.K. by radio that Sollum had been taken at 3:15 P.M. The assignment given on April 11, 1941 had thus been completed by April 13. On April 13, 1941 the *Knabe* advance unit had faced the enemy: approximately one (motorized) infantry battalion, 20 to 25 tanks and armored scout cars, three land batteries as well as the artillery of two or three ships. And the enemy had held absolute air superiority. About a quarter of all the motorcycles and their crews had been left in the desert, stuck in the sand on account of the roadless terrain.

"Only on the evening of April 13, 1941 did they make their way back, so that the attacks on Fort Capuzzo and Sollum had to be carried out with considerably decreased fighting strength. After Sollum was taken, the commander of the *Knabe* advance unit decided to switch to defensive action to secure the Bardia-Capuzzo-Sollum sector because of the fuel shortage and the threat of an enemy flank attack from the south.

"On the day after the taking of Sollum, frequent heavy fire from three or four land batteries and from naval artillery (cruisers and destroyers) began to fall on Sollum. This fire was continued with heavy support by the Royal Air Force in repeated bombing attacks against the entire position of the advance unit and Fort Capuzzo. After remarkably heavy artillery preparation, the enemy attempted, at 12:35 A.M. on April 15, 1941, to retake Sollum by shock-troop undertakings by selected units (about one column) against the heights. The enemy advance was repulsed, at great cost to the enemy, in a short battle. The enemy repeated this attempt after about an hour's artillery preparation from land and sea during the following night, this time with considerably stronger forces against the advance unit's position in Sollum. Here the British were armed with, among other things, long broad-bladed knives with their handles forming knuckledusters (and stamped "Made in New York"). This British counteroffensive was also repulsed, with major losses for the enemy. Eighteen British soldiers lay dead before just one Pak gun. Once again, Sollum remained firmly in the hands of the *Knabe* advance unit."

The motorcycle riflemen of the 3rd Company of Motorcycle Rifle Battalion 15 were the first German soldiers who entered the Land of the Pharaohs with weapon in hand. And the taking of Fort Capuzzo and Sollum, despite the toughest enemy resistance, meant the winning of an area of land that was to be of decisive importance to the coming operations.

Germany, spring 1940: The sergeant, motorcycle rifle column leader, hurries to his BMW R 12 sidecar after the first command: "Mount!"

Germany, spring 1940: Cross-country training. A motorcycle rifle group on heavy BMW R 12 1935-41, 20 HP, 750 cc SV units.

Germany, spring 1940: A heavy BMW R 12 1935-41, 20 HP, 750 cc SV unit crosses a brook. When crossing water (army instructions): "Select low gear! Drive slowly! Do not choke the motor!"

Germany, spring 1940: The heavy BMW R 12 1935-41, 20 HP, 750 cc SV machine during cross-country training. When driving downhill (army instructions): "Slow down!"

Germany, spring 1940: The cross-country driving training of a motorcycle rifle unit on the heavy BMW R 12 1935-38, 20 HP, 750 cc SV units. Getting out of a ditch (army instructions): "Lift forward and upward!"

Germany, spring 1940: The motorcycle rifle group on heavy BMW R 12 1935-41, 20 HP, 750 cc SV machines in marching order on the way to their cross-country combat training. The group leader is in the sidecar of the first motorcycle. On both sides of the road are PK cameramen, at left with an Ariflex 35 mm, at right with an Askania 35 mm on a tripod.

Upper Rhine, spring 1940: Before leaving for motorcycle rifle training. The heavy BMW R 12 unit still bears its civilian registration number.

Upper Rhine, spring 1940: Street-fighting training for the crew of a heavy BMW R 12 1935-41, 20 HP, 750 cc SV cycle. The riflemen are armed with 98 b carbines and wear 1935 model steel helmets as well as water-repellent motorcycle coats. The cycle bears a civilian registration number from the capital, Berlin. (The forward rifleman is left-handed.)

Upper Rhine, spring 1940: Cross-country training for a motorcycle rifle column, with a heavy BMW R 12 1935-41, 20 HP, 750 cc SV cycle. If the motorcycle has gotten stuck (army instructions): "Dismount! Lift out the front wheel! Push with first gear engaged!"

Upper Rhine, spring 1940: The motorcycle rifleman in the sidecar of the BMW R 12 with an MG 34 machine gun after cross-country training.

Western front, spring 1940: A heavy BMW Luftwaffe unit with an MG 30 drum-feed machine gun moves out.

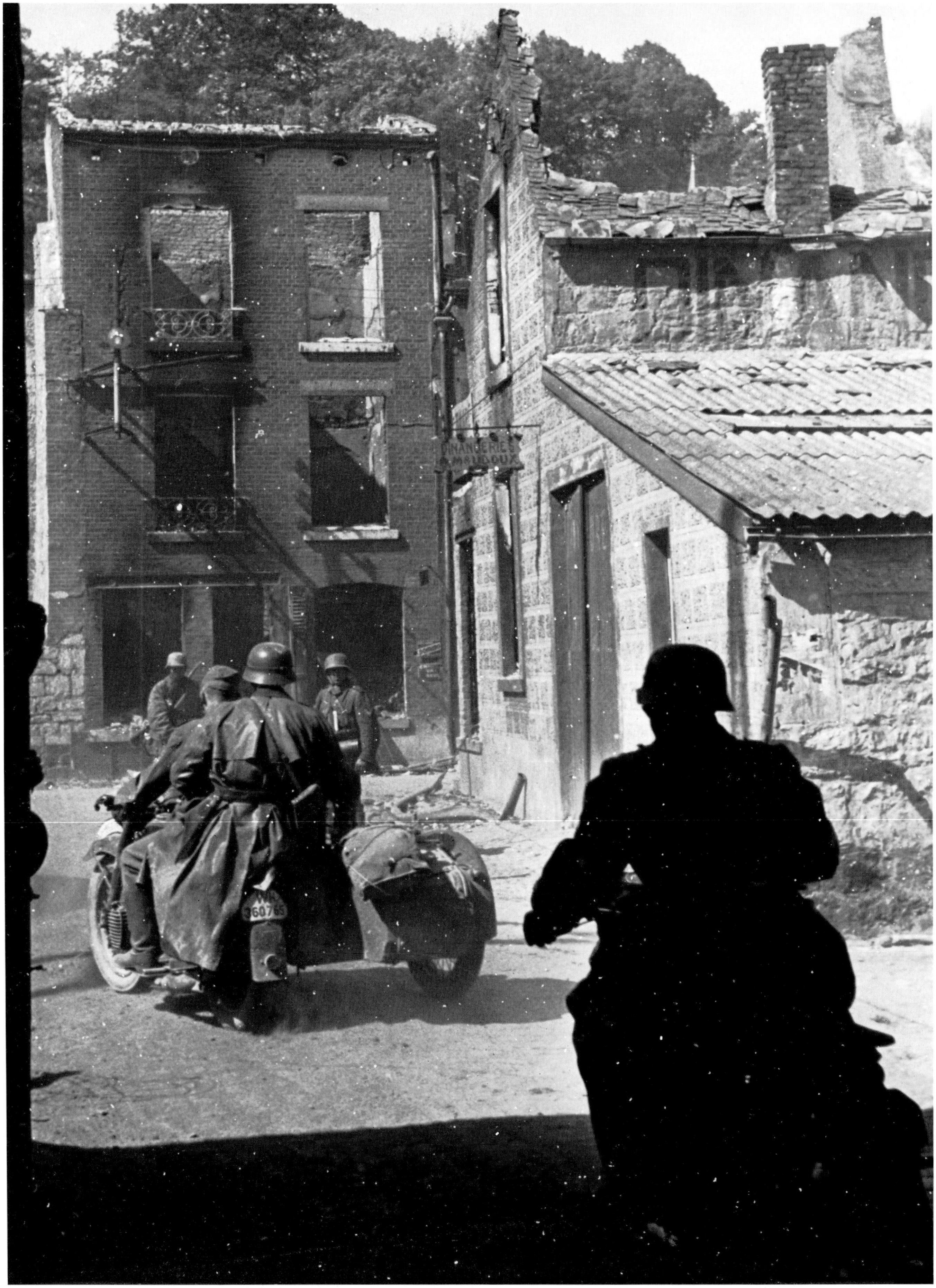

« France, May 1940: A motorcycle group on its way through Arras. On the heavy BMW cycle is the passenger, wearing a motorcycle coat and carrying a despatch pouch.

French campaign, summer 1940: A motorcycle messenger wearing the 1935 model steel helmet and the motorcycle driver's goggles made by Leitz. Over his water-repellent motorcycle coat he carries a gas-mask case. At left are his cooking utensils, at right his bullet pouch, both fastened to his belt.

French campaign, June 20, 1940: "Lyon taken in battle" says the headline of the army newspaper "Der Vormarsch." The crews of the BMW R 12 1935-41, 20 HP, 750 cc SV machines wear 1935 model steel helmets for which they have made rubber rings in which to stick camouflage materials, and water-repellent motorcycle coats. In the pouches hanging over their chests they carry anti-gas covers. A fuel can has been attached to the second seat of the front cycle. A cloth cover is on the headlight of the second cycle. In front, over his gas mask, the driver has a service flashlight, the so-called uniform flashlight with leather strap and buttonhole. The lamp itself is equipped with three small push-buttons that activate the sliding lenses and turn the normal white light into red, blue or green light.

French campaign, near Paris, June 1940: A motorcycle messenger of an infantry regiment, on a heavy BMW R 12, 750 cc SV cycle. The machine has bumpers to protect the cylinders in rough terrain.

French campaign, summer 1940: A motorcycle rifle advance unit at rest. A heavy BMW R 12 1935-41, 20 HP, 750 cc SV cycle. Between its masked headlight and handlebars are two Type 24 stick grenades with igniter 24 and detonator No. 8.

3

From 1934 on, the fading significance of cavalry was balanced by the rise of both armored vehicles and motorcycle rifle units. In 1935, after the full military strength of the Third Reich had been attained, the motorcycle rifle units assumed the role of the old mounted scouts. Independent motorcycle rifle units gradually were formed in the army corps or as fast-moving units of the Panzer divisions.

The motorized infantry divisions and the Panzer divisions were also given reconnaissance units. These consisted of the staff with an intelligence unit, an armored scout car company and three motorcycle rifle companies, and sometimes also a heavy company with light infantry guns, a motorized antitank platoon and an engineer platoon.

The fast-moving forces with their motorcycle rifle companies and motorized reconnaissance units proved themselves particularly in Poland in September of 1939 and during the western campaign in 1940. Shortly afterward, the rifle regiments of the motorized Waffen-SS divisions were given a so-called 17th company in addition to the usual 15th motorcycle rifle company, and the two became the reconnaissance unit of the regiment. The 15th company, and sometimes the 17th as well, originally consisted of three motorcycle rifle platoons and one heavy machine gun group. Only in the winter of 1941-42 were they given a complete heavy MG(-4) platoon, consisting of two heavy MG groups, each with two heavy machine guns. The 15th and 17th companies also acquired a grenade launcher group, each with two (8 cm) grenade launchers. The 15th motorcycle rifle company, strengthened with a Pak gun, a machine gun and an engineer platoon from the 13th, 14th and 16th companies, was usually, depending on the situation, given the assignment of being the advance unit assuring the steady advance of the regiment.

The 15th company, strengthened by parts of the 13th, 14th and 16th companies, could — on its own — not only provide reconnaissance but also make extensive advances, block roads and bridges and carry out flanking or surrounding moves. The 17th company generally provided reconnaissance for the regiment. When the advance unit, the strengthened 15th company, encountered the enemy, the spearhead began the combat. If the enemy turned out to be stronger than expected, the other motorcycle rifle platoons and the weapon groups subordinated to them went into battle. If the entire advance unit could not handle the enemy, then the lead battalion of the advancing regiment appeared on the battlefield, followed by the whole regiment. According to the service instructions then in force, here is how it looked:

Conduct and combat of the motorcycle rifle battalions
(according to the service instructions of the high command of the army, December 18, 1941)

Composition of the motorcycle rifle battalions Staff
with intelligence platoon
1 armored scout company
3 motorcycle rifle companies
1 heavy company
(consisting of)
gun platoon
antitank platoon
engineer platoon
supply trains
(consisting of)
service train I with supply unit
service train II with repair unit
pack train
commissary train

1. The motorcycle rifle battalion is the fastest and most mobile unit of the fast-moving forces. It carries out the tactical reconnaissance of the Panzer division and motorized infantry division and is capable of carrying out any infantry combat assignment.
2. With its large number of machine guns, the battalion has a high firepower, which along with its great mobility on the battlefield makes it a strong, fast and versatile combat unit, capable of obtaining clear information as to the enemy position during combat.
 A quick change between combat on foot and movement on vehicles is the most significant characteristic of its type of combat.
3. Its being equipped with motorcycles allows a quick advance on all roads and paths as well as across country in dry conditions. Long marches in bad weather (ice, snow), in unfavorable terrain (sand), in the dark or in fog demand much of the motorcycle riflemen's and their motorcycles' power and decrease their performance.
4. Motorcycle riflemen on their motorcycles offer a small target and can utilize the terrain quickly and efficiently. They are ready to fire quickly, can withdraw quickly from the enemy's view, and can cross areas under enemy artillery fire quickly. They are sensitive to concentrated enemy infantry fire. Enemy fire and unfavorable terrain can limit the movement of the unit with vehicles on the battlefield and compel all or part of the battalion to dismount. Therefore riflemen on armored vehicles are better suited to accompany a tank attack than are motorcycle riflemen.
5. The main assignment of the battalion is to carry out and protect the tactical reconnaissance of the division. For this, it should be deployed in as unified form as possible. Combat assignments often have to be combined with the reconnaissance assignment when the enemy position or lack of other forces compel it. The battalion is particularly capable of carrying out the following assignments:
a. Deployment as an advance unit,
b. Movement against the enemy's flank or rear for the purpose of surprise attacks,
c. Pursuit, especially for the purpose of overtaking,
d. Protection during rest and concealment of the movements of motorized forces.

In defense, the battalion can be deployed to protect open flanks. The best protection is extensive reconnaissance. Use of the unified battalion as a mobile reserve is generally considered only when reconnaissance assignments of the battalion can be dispensed with for the time being.

6. To carry out large independent combat assignments, the battalion can be strengthened by antitank forces, engineers and particularly artillery, depending on the situation.

Means of Warfare

7. The means correspond to those of the infantry battalion. But the motorcycle rifle battalion possesses a considerably higher firepower and, through its armored reconnaissance company, a faster means of armored reconnaissance with a large range.

8. The close relationship between the motorcycle crew and its vehicle caused by the constant change between movement on the motorcycle and combat on foot makes the motorcycle a particularly important means of warfare. The speed and mobility of the motorcycle are always to be utilized to the fullest.

9. The performance capability of the individual motorcycle (messenger driver) amounts to a speed of 70 to 80 kph on good roads. On the bad roads and paths, across country, in bad weather conditions, with limited visibility and when riding with groups, the speed is decreased significantly. The attachment of a machine gun on the bow of the sidecar makes firing from a moving or halted motorcycle possible and thus provides constant readiness to fire on unexpectedly appearing ground targets.

Influence of the Terrain on Mobility and Type of Combat

10. The battalion's speed of mobility, reconnaissance and combat are highly dependent on the condition and closeness of the available network of roads and paths and changes to them caused by weather conditions. Early orientation over the paths and weather conditions in the area of action, as well as timely reconnaissance, are thus of particular importance. Lack of paths, combined with long periods of rain, can destroy the usefulness of motorcycle rifle units.

Reconnaissance

11. Reconnaissance should provide a picture of the enemy as quickly, as completely and as reliably as possible. Its results form the most important basis for the division's procedures and for the utilization of weapon effect.

12. Within the framework of reconnaissance activity, the reconnaissance of enemy armored and antitank forces and the knowledge of roads and terrains — particularly of barricades and traps — are of particular importance. Through early familiarization with the roads and terrain, the motorcycle rifle battalion can provide the leadership with the basic information necessary for the movement and deployment of the division.

Assaults

13. The speed and mobility of the battalion must be utilized fully in the conduct of combat. Especially important for this purpose: purposeful conduct, concerted formation of focal points, quick decisions, quick and concise giving of orders (generally single commands) and forward-looking management of the back-line forces, especially fuel supply.

14. Establishment and securing of vehicle echelons, their timely provision and readiness for the dismounted troops and placing the troops on the battlefield are usually ordered by the companies.

15. So as to avoid having too many vehicles on the battlefield or limiting the mobility of other units, it can be advantageous for the battalion to direct the companies to move

their vehicle echelons. In special cases, especially when crossing rivers, it is necessary for the battalion to command the vehicles.

Tactical Reconnaissance and Messenger Service

16. The assignment of reconnaissance must be given to the battalion early enough so that it can give the scouting troops the necessary advantage (at least one hour) and gain sufficient ground before the mass of the division (two to three hours, if possible).

17. When far away from the enemy, generally only a few scouting troops are sent out at first. The assignment of this first wave of reconnaissance is to locate and identify the enemy in a particular area or make contact with the enemy reported by aerial reconnaissance. For this purpose it is generally sufficient to seek out the most important roads and transit lines as well as the localities. The subsequent assignment is to determine the position, the composition and the activity of a spotted enemy. If the first action of the scouting troops is not sufficient to obtain adequate information as to the enemy's situation in the area being reconnoitered and to fulfill the assignments given by the division, additional scouting troops must be sent out with clearly defined single assignments. If the battalion receives new reconnaissance assignments from the division in the course of its action, then the deployment of new scouting troops is usually more efficient than the transmission by radio of new assignments to scouting troops located near the enemy, since the requirements for adequate issuing of commands are not provided. This is especially the case with assignments that direct troops in a completely new direction. On the other hand, previous assignments to scouting troops can be enhanced or extended to further assignments by radio. The sending out of new scouting troops is likewise necessary when reports from scouting troops are lacking and the presumption arises that they have been lost through contact with the enemy. The battalion commander must therefore always retain sufficient reserves for reconnaissance.

18. Contact between the scouting troops and the battalion commander is maintained by radio. When radio silence is ordered, reports can be delivered by detailed messenger vehicles (motorcycle messengers or light armored scout cars). Transferring reports to a fixed radio position known to the enemy is also a possibility.

19. When marching against the enemy, the battalion forms a reconnaissance reserve, a report collecting position and a holding position for the scouting troops. If the situation is not yet clear, the division generally orders the battalion to move forward from sector to sector. If the battalion is left to select the sectors, the length of the advance depends on the closeness of the enemy and the nature and cover of the terrain. As the enemy comes nearer, the advances are to be shortened. Apparently enemy-free terrain is to be traversed quickly.

20. To secure the march, the battalion is formed into an advance guard and main body. One armored spearhead is to be positioned at the head of the advance guard and one at the head of the main body. It is practical to send out a scouting troop before the spearhead with a lead of about ten minutes.

21. The strength of the advance guard depends on the assignment, situation, terrain and the strength of the main body to be secured. It generally consists of the armored reconnaissance company and one motorcycle rifle

company. The deployment of the antitank forces and parts of the engineer column is the rule. Intelligence troops are to be positioned far forward. The main body generally follows the advance guard at an interval of about ten minutes. Situation, terrain and weather can necessitate a different composition and different intervals.

Operations of the Motorcycle Rifle Company
(according to the service instructions of the high command of the army of March 16, 1941)

Composition of the Motorcycle Rifle Company
Company leader with company troop
3 motorcycle rifle platoons
(18 light machine guns, 3 antitank boxes, 3 light grenade launchers)
1 heavy machine gun group (2 heavy machine guns)
Service train
Vehicle repair troop
Pack train

Nature, Assignments and Composition of the Company
1. The motorcycle rifle company combines speed, mobility and high firepower. With favorable ground conditions it can also carry out the movement of vehicles off roads and paths.
2. Quick change between fighting on foot and quick mobility on vehicles is the significant characteristic of its conduct of combat.
3. Motorcycle riflemen are particularly capable of the following assignments: Taking quick possession of important territory, keeping narrows open or closing them in the face of the enemy or in combat with weak enemy forces, surprise attacks against the enemy's flanks and rear, pursuit, securing and concealing the movements of motorized forces.
4. The motorcycle rifle company generally fights as part of the motorcycle rifle battalion, armored reconnaissance unit or motorized reconnaissance unit.
5. The high number of machine guns and the possession of steep-fire weapons also allows it to carry out independent combat assignments. Generally it must then be strengthened with heavy and armor-piercing weapons.
6. Within the reconnaissance unit it will often have the assignment of making reconnaissance possible through combat in unified action. Parts of the company can be utilized to strengthen or concentrate the reconnaissance.

The Group

Composition, equipment and assignments
7. The motorcycle rifle group consists of: group leader, 7 riflemen (two of them with a machine gun), 4 motorcycles with sidecars.

The Manner of Assault

Advancing against the enemy and dismounting for assaults
8. In order to move forward quickly, the group will travel on roads if possible. The closer they come to the enemy, the more they must accommodate themselves to the terrain when advancing. The motorcycles' quickness, mobility and ability to handle the terrain are to be utilized fully.
9. If there is no further possibility of approaching the enemy on vehicles under

cover, or if terrain and enemy fire compel, they are to dismount.

10. For combat, the riflemen sit until commanded "Equipment free — dismount for combat!" The equipment is detached. The riflemen prepare to fight and form, if nothing else is ordered, the "rifle row."

11. In a sudden confrontation with the enemy, the riflemen take up positions immediately after dismounting to be able to commence firing.

12. The motorcycle echelon is to take cover quickly after dismounting. Their location in the terrain is to be ordered by the group leader by words or pointing. If this is not possible, the motorcycle echelon leader is to lead the echelon to cover independently.

13. If they dismount in enemy fire, the motorcycles are to be moved out of the fire as quickly as possible.

14. When the motorcycle rifle group advances, a machine gun generally forms the spearhead. The longer the group follows this machine gun in a thin, deep formation, the longer machine guns farther back in the group can shoot past it.

The Group as a Scout Group and a Spearhead Group

15. If the group is utilized as a scout group or a spearhead group, it generally travels in marching order. The group leader orders the succession of marching. According to the assignment and the situation, he will often detail a machine gun cycle as the leading vehicle.

16. In order to lead as quickly and with as great mobility as possible, the group leader always rides on the leading motorcycle. Usually he will ride on the back seat, since he has a better field of vision from there than from the sidecar.

17. In order to limit the effect of enemy weapons, the motorcycles generally travel at long intervals (sight intervals). In localities and in vision-blocking terrain the intervals are shortened.

18. The group is generally assigned one or more motorcycle messengers to deliver reports.

19. The machine guns are mounted on the sidecars with barrels in place, ready to fire. The protective covers are to be removed at the right time. In unclear situations rifles and pistols are to be carried ready to fire.

20. Mine barrages are to be looked for, particularly at narrows, bridges and entries to localities. In case barrages cannot be removed quickly by the group, they are to be circumvented, so as to lose no time. They are to be made visible, and their locations are to be reported immediately.

The Patrol

21. Enemy-free terrain is to be traversed quickly. If contact with the enemy is likely, the patrol is to move ahead according to visibility in jumps from one observation point to the next. If the situation requires, the patrol is to provide its own fire cover while advancing.

22. As long as visibility and enemy position allow, the patrol is to remain mounted. Unnoticed and covered approach to the enemy, observation from lookout positions (church steeples, trees, etc.), location of explosive charges on bridges and barricades or the effect of enemy fire can make dismounting necessary. The motorcycles are then to be kept nearby. They are to be brought up quickly on signal.

23. If the purpose of the reconnaissance is fulfilled by making contact with the enemy, then the patrol is to stay close to the enemy

unless otherwise ordered. Contact with the enemy is not to be lost. If the patrol is to keep contact with the enemy at night, they will often keep watch over the enemy on foot through reconnaissance from cover (woods, farms, etc.).

The Spearhead
24. In case of a surprise encounter with the enemy, the leading machine gun cycle is to commence firing while in motion, so as to give the group fire protection while dismounting and taking up positions.
25. Dismounting and taking up positions must be done lightning-fast. The motorcycles are to be placed under cover quickly to clear the road.
26. In case of an encounter with enemy armored vehicles, the following troop is to be warned immediately by motorcycle messenger, light signals or other understandable signals.

The Group as Field Sentries
27. If nothing else is ordered, the motorcycles are to remain under cover in the vicinity of the securing group. Replacements and changes of position can be carried out quickly with the help of the motorcycles.

The Heavy Machine Gun Group
28. The heavy machine gun group consists of: 1 heavy machine gun group leader, 1 messenger, 1 rangetaker, 2 machine gun crews, each with 1 gun leader and 5 gunners, 10 drivers, 10 motorcycles with sidecars (4 per gun).
29. The heavy machine gun group is, on account of its firepower, a significant weapon in the hands of the company leader to build focal points of fire and thus to influence the conduct of the battle.

30. The nearer the company comes to the enemy, and the more necessary support is for the breakthrough of the group, the closer the heavy machine guns must remain to the parts of the company fighting in front.
31. If the breakthrough succeeds, the heavy machine gun group independently follows the parts of the company that have broken through, in order to be available soon for new assignments. These consist of supporting the company in fighting in the depth zone by repelling enemy counterattacks or by halting enemy flank attacks, by fire on individual enemy nests and by holding back or wiping out troublesome targets in the depth of the main enemy battlefield. During the breakthrough and after it has succeeded, the heavy machine gun group, especially in fighting off enemy counterattacks, must often act on its own judgment.

The Platoon Headquarters
32. The platoon headquarters consists of: 1 platoon leader, 1 messenger (simultaneously bugler), 2 motorcycle messengers, 2 truck drivers, 1 stretcher carrier, 1 medium personnel carrier with equipment case, 2 motorcycles, 1 fighting vehicle.

The Company Headquarters
33. The company headquarters consists of: 1 company leader, 1 driver of the company's vehicle echelon, 4 messengers (1 simultaneously bugler), 1 simultaneously shear telescope carrier), 4 motorcycle messengers, 2 truck drivers, 1 stretcher bearer, 1 medium personnel carrier with equipment case, 1 fighting vehicle, 4 motorcycles (2 of them with sidecars).

The Platoon
34. The motorcycle rifle platoon consists of: 1

platoon leader, 1 platoon troop, 3 light machine gun groups, 1 light grenade launcher troop.

The Platoon and the Company in Combat

35. The motorcycle is a fighting implement of the motorcycle rifleman. The company leader and his junior leader must always be sure to utilize the speed, mobility and cross-country capability of the motorcycles fully. Only when an assignment cannot be carried out on motor vehicles is the decision to dismount justified.

Marching

36. The weapons are to be protected from dust. In transition to marching against the enemy, the windshields of the motorcycles are to be folded down and camouflaged. The company leader is to decide when the riflemen are to remove their protective coats. He also orders when to take the weapons out of their protective covers, when to load, and whether the machine guns on the sidecars are to be made ready to fire.

37. If a machine gun cycle breaks down on the march, the machine gun is to be taken along by the next motorcycle.

38. Motorcycle drivers are always to wear goggles when on the march. The goggles, when taken off, are to be worn around the neck.

39. Protection against coldness is especially important for motorcycle riflemen. On longer marches, short stops must often be made for warming.

40. For the march, a top speed (top speed of the leading motorcycle) can be ordered. In the daytime this will generally be 35 to 45 kph. When the assignment and situation allow, the top speed can be raised in favorable road and visibility conditions. Snow, ice, steep grades, etc. can decrease the speed considerably.

41. The average daily distance covered on a march without enemy contact on traffic-free roads amounts to 250 kilometers. It can be increased by changing drivers. At night and with poor visibility, the marching distance will be decreased according to the visibility and the ordered stage of illumination.

42. Long driving in lower gears is damaging to the motors of the motorcycles and considerably increases the fuel consumption.

43. During the march, the cohesion of the company must not be lost. This requires particular attention at places on the road where one cannot see far ahead, in towns, in dust and limited visibility. If the marching column breaks as a result of a suddenly appearing obstacle, the platoon leaders are to wait until their platoons are together and follow along. To maintain and reestablish connections, the company leader and the platoon leaders are to use motorcycle messengers.

44. When stopping, one should turn sharply to the roadside, if necessary using the sidewalk. Generally the vehicles do not close ranks, but remain some distance (intervals of at least 10 meters) apart. Aircraft covers are to be used if available. Traffic control posts are to be set up to let other vehicles and columns pass by. Connections between all parts of the company are especially important for fast transmission of orders and smooth resumption of the march.

45. Rest stops are generally made every 4 to 5 hours. If maintenance is carried out during the rest, its length should be at least 2 to 3 hours.

46. At the right time before every rest stop, a reconnaissance troop is to be sent out to find a suitable resting place with possibilities for

turning aside, entrance and exit routes, and to direct the arriving troops. Woods, groves and cover of all kinds off the advance route are suitable for rest stops. If a stop has to be made on the route of march, the road is to be left fully open.

47. When a company marches against the enemy alone or as a spearhead company, it secures itself by separating a spearhead.

48. Terrain and visibility conditions are crucial in determining the interval between the company and the spearhead platoon, as well as the intervals within the spearhead. An interval of 3 to 4 minutes between the company and spearhead platoon, and of 1 to 2 minutes (sight interval) between the spearhead platoon and company, can be regarded as standard. 49. If the spearhead encounters resistance from the enemy, the company must stop and take cover at the right time in order to avoid running into the enemy and to maintain its mobility.

50. If the company travels as a rearguard company, it is to secure itself by forming a rearguard.

Overnight Stops

51. Concealing the vehicles from enemy air reconnaissance must be done very carefully. Garages, barns and sheds are to be used for making camp. Parking places may not be set up.

Deploying

52. Fixed deploying formations for motorcycle rifle companies do not exist. The width and depth of a formation depends on the enemy, the terrain and the intended course of action, and is to be commanded according to the situation.

53. The point in time at which deploying on motorcycles gives way to advancing on foot depends on the assignment, enemy action and terrain. The speed of the vehicles is always to be utilized for advancing as long as the effect of enemy fire and the terrain allow. Early dismounting to approach the enemy without being seen can be effective.

Attack while in Motion

54. The vehicle echelon is to stay as near as possible to the dismounted troops in combat, in order to be able to make the company or individual platoons mobile again quickly. In a progressive attack, the vehicle echelon is therefore to follow the fighting troops from cover to cover.

Cooperation with Armored Forces

55. Speed and mobility on the battlefield allow the motorcycle rifle company to take quick advantage of tank advances. Here it can be the motorcycle riflemen's assignment to cooperate closely with the tanks or ahead of them in taking possession of important, strategically decisive points and sectors of the terrain.

56. If the company is to follow the tank attack, then it is to remain mounted as long as the terrain and enemy action allow. It is to keep far enough behind the tanks so that it is not exposed to fire directed at them. It should form in depth and width so that it can make extensive use of available roads, terrain that offers cover, and low-fire areas for its movements. Compactness is to be avoided. Every crippling of the enemy achieved by fire from the tanks is to be utilized by the company for fast advancing.

Pursuit

57. For pursuit, the company leader must utilize the high speed and mobility of the vehicles. He must make every effort to stay on

the enemy's heels and to cut off his retreat by getting ahead of him where he can.

Defense

58. In defense, it is possible for the motorcycle rifle company, what with its mobility and possession of numerous machine guns, to defend itself over a broad front as an infantry company. Parts of the company that remain mobile can be deployed quickly to strengthen the defense at a danger point, to secure a threatened flank or to make a counterattack. The location of the reserves is to be chosen so that they can be deployed in various directions. Reconnaissance for this purpose is to be carried out at the right time by the leader of the reserves.

Breaking off Combat

59. If the company is utilized as a rearguard, it is to use every opportunity to strike at the enemy with fire on front and flank and slow him with counterattacks.

60. Disengagement from the enemy is to be made easier through darkness or artificial fog.

61. The speed of the vehicles makes it possible to reestablish contact quickly with the withdrawing troops, even over considerable distances, or to gain time and space to set up defenses at another position.

Assaults on Rivers

62. Motorcycles can be taken across rivers on pontoons, pontoon rafts, motorcycle riflemen's footbridges or improvised means. The motorcycle rifle company is therefore able to ferry a part of its motorcycles across without having to await the building of bridges. By utilizing the speed of the ferried vehicles, the company can quickly form an extended bridgehead.

63. The vehicle echelon must be held far back until crossing, so as to avoid a concentration of vehicles in the vicinity of the crossing points. The fast forward movement of individual cycles or groups for crossing is to be directed by clear communication. After reaching the far shore, the vehicles are to leave the crossing points at once and to take up the specified positions by the quickest route.

Assault Against a Strengthened Position

64. In combat against a position strengthened by expanding installations, the speed and mobility of the motorcycle rifle company cannot be utilized at first. The company is thus generally held back in order to advance quickly after the creation of gaps and attack the rear of the enemy position.

Assaults in Darkness and in Fog

65. In combat in darkness or in fog, a close connection with the vehicle echelon is to be maintained. It must stay close to the fighting troops. Its positioning must be done so that even at daybreak or lifting of the fog, the vehicles can be removed from the enemy's view. Its securing is to be strengthened.

Assaults for Towns and Woods

66. Town and forest combat is often costly. It can often be avoided by moving sideward and going around the town or forest. It can, though, be the assignment of the motorcycle riflemen to clean the enemy out of towns and woods that have been gone around by their own armored troops in combat. The motorcycle rifle company is generally strengthened with heavy weapons for this purpose.

67. The vehicles are not concentrated in the vehicle echelon of the company but follow close behind their platoons or groups from

cover to cover when the situation allows. The fighting troops thus have the possibility of being able to remount quickly.

Antitank Defense
68. In encounters with enemy tanks while on the march, the company is to clear the road in the direction indicated at the company leader's command or signal and seek the nearest covering terrain as fast as possible.
69. At close and very close ranges, firing heavy machine gun ammunition at the observation slits and loopholes promises good results from flying lead as well as through the effect on the crew's morale. Concentric charges of hand grenades should be prepared in advance and taken along.
70. In combat during an enemy tank attack, the company fights the enemy infantry attacking with or behind the tank forces to separate them from their tanks. At close ranges the company participates with all available means in fighting the enemy tanks. It withdraws from direct tank attacks by seeking cover. Even when enemy tanks have broken through its own front, the fight against the enemy infantry is to be carried on with all strength.

The Pack Train
71. The pack train consists of: 1 unit pay clerk, leader (on a motorcycle), one truck driver, 1 shoemaker, 1 tailor, both also truck attendants, 1 truck (1.5-ton) for baggage.
72. The pack train serves to carry baggage as well as supplies of clothing and materials for repairing clothing and equipment not needed in combat.

The Supply Service
73. Supplying the fighting vehicles with ammunition, fuel and rations from the service train or other supply points is ordered by the company leader. The platoon leaders are to report the state of their supplies to him at the right time.
74. The ammunition supply for the fighting units is brought by the supply truck or the reserves who come on motorcycles. To bring it forward, drivers on foot or on motorcycles can be used.
75. Fuel is generally supplied from canisters carried on the supply truck. Refueling can frequently be done at a stationary gasoline pump or from an available tank truck.
76. The supply unit is generally supplied from the supply issuing station of the next higher unit. In an emergency, supplies can be brought in from the surrounding country by the company on command. The supply train leader and the pay clerk are responsible for receiving supplies on time.
77. Supplying the fighting units of the company with rations via food carriers must be prepared and secured. Drivers can be detailed to serve as food carriers.

Motorized Messenger Service
(according to the instructions of the high command of the army of November 13, 1941)
1. On the march and in combat, the motorcycle messenger is an important means of communication.
2. For reconnaissance assignments, the motorcycle messenger must have infantry-engineer-technical training and instructed as to the number, type and performance capability of the vehicles in his regiment, their weights and track widths.
3. The giver of the assignment must tell the motorcycle messenger when handing him a message to be delivered:
(a) Where and to whom the message goes,
(b) Where the recipient is to be found,

(c) What is to be done if the recipient is not found at the specified place,
(d) What the contents of the message are,
(e) What to do after delivering the message.

The motorcycle driver must:
(a) Repeat the message after being given the assignment and, if necessary, clear up anything unclear,
(b) Destroy the message at the right time (for example, dip it in a fuel tank and burn it, etc.) if it is in danger of falling into enemy hands,
(c) Drive around towns, if possible, if there is danger of being attacked there,
(d) Reduce his speed on unfavorable terrain (the best message is useless if it accidentally arrives too late or not at all),
(e) Make use of his own troop's radio and telephone connections if necessary,
(f) Quickly stop passing troops and refuel if he is out of fuel,
(g) In the face of immediate danger, call out the gist of the message to the leaders of securing troops or troop units,
(h) Ask unabashedly for the whereabouts of the leader he seeks,
(i) Not betray command posts to the enemy by driving up to them, but halt at the level of the parked vehicles and dismount,
(k) Shut off the motor at radio and command posts,
(l) Interrupt commands vocally to deliver important messages,
(m) Ask for further orders after delivering the message.

4. To deliver messages and orders while underway, the motorcycle messenger identifies himself at the right time by raising his hand or calling, "Message for . . . !" Written messages are handed over while in motion, oral messages are called out. Messengers in sidecar motorcycles can also deliver the message by climbing into the leader's vehicle if this is advisable. Traffic in the opposite direction must not be held up. After delivering the message, the motorcycle driver is to clear the road immediately.

5. If the recipient of the message is riding in the opposite direction, the motorcycle messenger turns at the right time, so as to be able to ride beside the recipient's vehicle while giving the message.

6. The motorcycle messenger is to utilize waiting times and lulls in battle to maintain and refuel his motorcycle, the regular use of which requires particularly conscientious service.

The practical application of the motorcycle riflemen's service instructions was experienced by NCO Klaiber of Panzer Regiment 39 of the 17th Panzer Division when he and his small motorcycle group moved quickly to capture the Russian town of Lipka, not far from Sjenno.

Hans Klaiber: "The heavy BMW motors roared and dust rose high. The three motorcycles of my advance group got underway. I felt how the wind blew in my face. In this night of July 8, 1941 oppressively sultry weather lay over the land between the Düna and the Dniepr. Out on the road, a column was also moving in the direction of Lipka. I pulled down my goggles and looked down the road through the night glass. There were horse-drawn wagons that moved along the right side of the road.

"I grasped the butt of the machine gun and stared ahead again. Were they refugees, or was it a retreating Soviet unit that was trying to reach Sjenno? 'Come on! Move out!' I called to the driver. *Obergefreite* Decker accelerated. The two other cycles of the advance group followed us. To the left and right of the road, the bushes whizzed past us. Soon we caught up with the rumbling wagons, and at the same moment I knew we had a Soviet column before us. In a moment we on our cycles were past the wagons, in which exhausted Red Army men were crouching.

"Then we caught up with a marching column that was dragging itself along. I grasped the butt of the machine gun on the swinging arm of the sidecar and kept the figures we were passing in view. But nothing happened, and soon we had the column behind us too. On both sides of the roadway, fallow land stretched out; the bushes and telegraph poles flitted past. Presumably the Russians had taken us for their own men. Now we could see the outlines of small houses standing out against the sky. That must be Lipka!

"'Go down to forty!' I called to the driver. The motors became quieter and the cycles moved ahead slowly. Beside the road lay wrecked vehicles; apparently Soviet columns were surprised here by low-flying planes and shot up. Seconds later our advance group reached the town. Nobody detained us. On the street we drove along, gun pyramids stood before the houses, and Red Army men lay here and there, sleeping.

"At every moment I expected to be shot. We stayed in the middle of the street to keep as far away from the sleeping Russians as possible. The two other cycles followed us at short intervals. We knew that only boldness could help now. The cycles rumbled down the street one after another, past parked trucks. Here too, not a sentry was to be seen, only sleeping soldiers behind the rolled-down windows of the truck cabs.

"Now we rolled across the market place with the Lenin monument in the middle. 'Drive around the square!' I signaled to the driver. 'Maybe we'll find a side street we can escape through!' 'I'd like to get my hands on the idiot who told the Old Man this place was free of enemies!' Decker growled and pulled the handlebars to the right; behind us, the two other cycles turned off.

"Then a Soviet car roared out of a side street. A Russian officer jumped out of the car and ran toward our motorcycle. I saw him raise his machine pistol and reacted at once, swinging the barrel of my machine gun and firing. The salvos broke the stillness of the night. The officer fell in a heap. In the same moment the car caught fire and went up in flames, hit by a salvo from the cycle behind me.

"'Fighting curve!' I shouted, tore the machine gun out of its mount and tossed it to Schneider, who had jumped off the back seat. Then the second cycle roared up. I rushed toward it and shouted to the driver, 'To the next street! Block it! Don't let anybody through!' The machine roared away, crossed the market place and formed a fighting curve there too. The passengers put the machine gun in position and secured the street while the driver went on to park the machine in a safe place, a niche in the wall. 'To the street after that!' I shouted to the driver of the third cycle, and seconds later the mouth of that street was also blocked and the cycle parked in a driveway.

"I ran with Schneider to the mouth of the street at which we had come into the square, pulled out my flare pistol and fired. The flare cartridge rose high, hissing. Now the company knew that their advance group had gotten into the town. *Obergefreite* Decker threw himself on the ground with the machine gun. 'As soon as anything moves, shoot!' I ordered him and rushed to the next street corner. There was the second cycle's crew with the barrel of their machine gun pointed at the street. 'Don't let them come near!' I called and ran over to the third street corner. There were *Gefreite* Müller and two riflemen behind their machine gun.

"The Soviet car was still burning. The buildings around the market place threw restlessly fluttering shadows on the paving stones. Suddenly Decker's machine gun began to bark. I heard cries from the street leading in from the west. Decker covered the street with sweeping fire, and I saw the Red Army men rushing forward into the bursts of fire. Others pressed themselves against the walls of houses and returned fire.

"If the company did not arrive soon, we were all lost. Out of the street where the Russians were attacking came bullets, rattling against the wreckage of the Soviet car. Now Müller's machine gun started to fire from the street to the east. Only seconds later there was gunfire at the northern street corner too, where Hancke and his two comrades were.

"Now the Russians attacked, and within the shortest time the battle situation had changed. From all sides shots hammered and barked, people shouted and hand grenades detonated. Tracer bullets shot through the night. One of the houses on the northern edge of the square caught fire, and shortly afterward, a building on the west side went up in flames too. At this moment two other houses on the south side of the market place also began to burn. 'Look!' Decker suddenly shouted. A green ball of light was in the sky. That was our company! We could hear the hammering of German automatic weapons clearly from the west now.

"The Russians kept us under fire from the houses. On the street a few vehicles burned brightly, and I saw bodies lying motionless in front of Hancke's machine gun. Suddenly we heard the sound of an alarm whistle. In seconds the Red Army men rushed out of the houses and ran to the mouth of the street. Hancke's machine gun barked. Then it was quiet again. But meanwhile the corner house near Hancke had caught fire. Tongues of flame rose from the thatched roof, and cascades of sparks swirled through the air.

"I took the men back a few meters, warned them not to waste ammunition and ran back to Decker and Schneider. From the entrance to the town one could clearly hear the droning of motors, the hammering of German machine guns and the explosions of hand grenades now.

" Over the market place hung clouds of sparks that made us feel as if glowing snow were falling from the sky. A strong enemy colony went toward the street corner, in range. I waited a second, then ordered: 'Fire at will!' The Russian attack wave came to a sudden stop and broke up. Then a Kfz. 17 appeared and chugged past the burning column of Soviet vehicles. The noise of battle was silenced. Cycle after cycle hurried in and drove onto the market place. Scarcely fifteen minutes later we were sitting on our motorcycles again and charging off toward Sjenno."

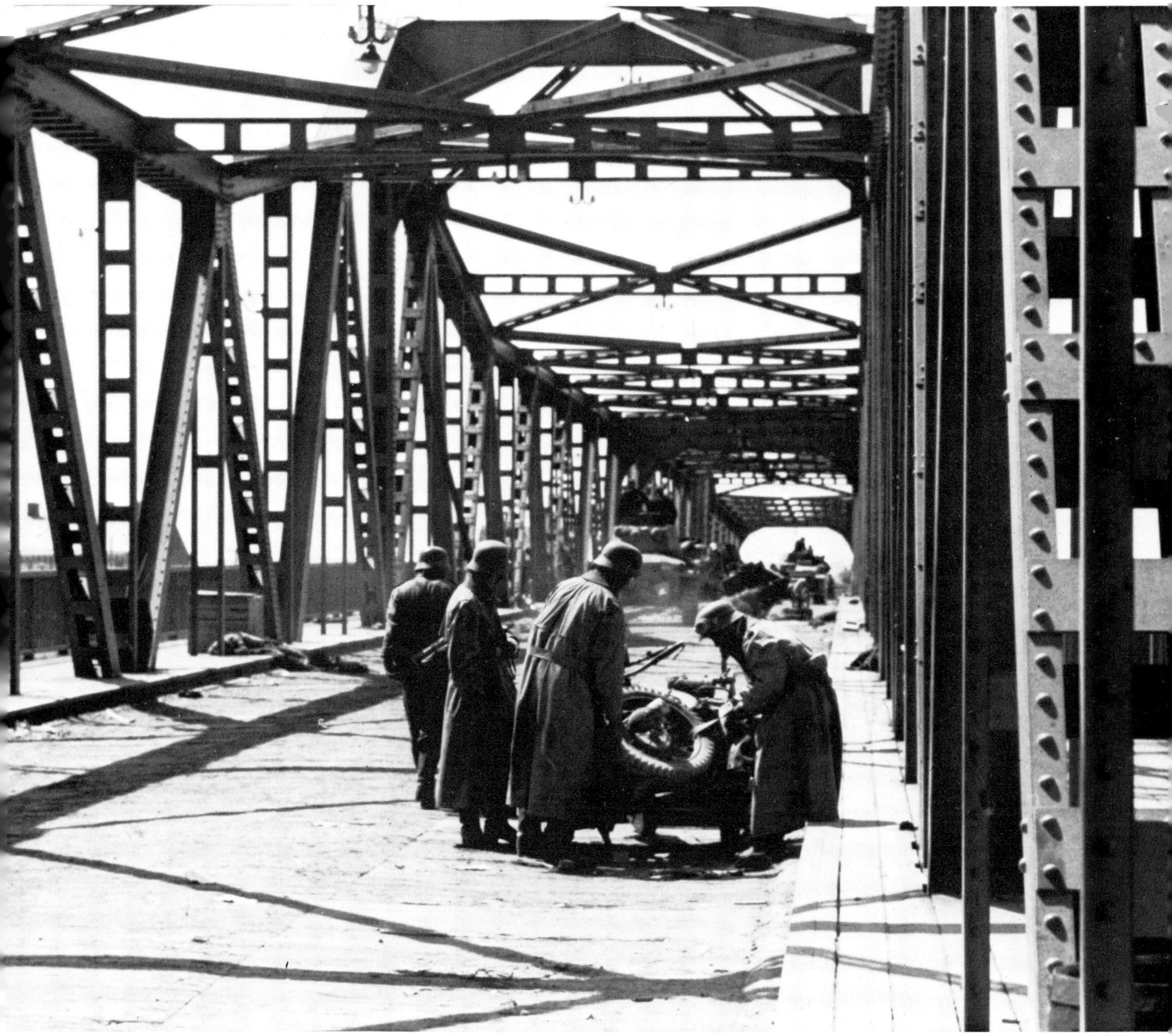

French campaign, summer 1940: A flat tire during the advance. The crew of a BMW machine changes a wheel.

French campaign, summer 1940: A campfire on the Loire. A motorcycle messenger in front.

France, summer 1940: A motorcycle messenger platoon of a Luftwaffe staff. At left a BMW R 35, 350 cc OHV medium cycle, and two R 4, 400 cc OHV Reichswehr medium off-road sport cycles (known as "Springers").

Germany, at the edge of an airfield, summer 1940: An *Obergefreite* of the Luftwaffe takes a short siesta in the sidecar of a BMW R 12 1935-41, 20 HP, 750 cc SV cycle.

Poland, spring 1941: The motorcycle rifle group of a motorized military police unit on the march. In front is a heavy BMW R 12 1935-41, 20 HP, 750 cc SV cycle.

Rheims, France, spring 1941: A motorcycle rifle unit prepares for combat drill in the country.

Near Rheims, France, spring 1941: Combat drill in the country. The captain dictates an order to the company troop leader for the motorcycle messenger. The cycle is camouflaged to avoid being seen by the enemy.

Near Rheims, France, spring 1941: Combat drill in the country. A motorcycle messenger with his heavy BMW R 12 1935-41, 20 HP, 750 cc SV machine after delivering a command. The officer is orienting himself by means of a map.

Near Rheims, spring 1941: The motorcycle messenger on his heavy BMW R 12 1935-41, 20 HP, 750 cc SV machine underway to deliver a command in rough country.

Near Rheims, spring 1941: Combat drill in the country. In front is a motorcycle rifleman on a medium BMW R 5 1936-37, 24 HP, 500 cc OHV cycle. On his 1935 model helmet he is using his musette bag strap to hold camouflage material.

Germany, spring 1941: The mechanics in a repair shop at work on a BMW R 4 1932-36, 12 HP, 400 cc OHV off-road sport Reichswehr "Springer" medium machine. At right is a DKW.

Germany, spring 1941: In front of a repair shop, a mechanic is checking the front fork of a BMW R 4 1932-36 12 HP, 400 cc OHV off-road sport Reichswehr "Springer" medium cycle.

France, spring 1941: Riding cross-country on a heavy BMW R 12 1935-41 20 HP, 750 cc SV cycle. The two soldiers wear motorcycle coats and 1935 model steel helmets. At left the machine-gun belt hangs out (very much against regulations) over the hull of the sidecar.

France, spring 1941: On the march across the fields. BMW machines during cross-country drill.

France, spring 1941: Motorcycle riflemen about to attack a town. The cycles are immediately driven into full cover (drill). In the center is a heavy BMW R 12 1935-38, 20 HP, 750 cc SV machine.

France, spring 1941: Combat around a village. The motorcycle rifle platoon during a fire-fight drill in moving against enemy positions.

France, spring 1941: The motorcycle riflemen during off-road drill. The forward heavy BMW machine is lifted over a stone wall. On the sidecar of the rear BMW is an MG 34 machine gun.

France, spring 1941: They did it! The heavy BMW machine rolled over a stone wall.

North Africa, Capuzzo-Sollum area, mid-April 1941: The *Knabe* advance unit on the march in their heavy BMW R 12 1935-41, 20 HP, 750 cc SV machines.

North-Africa, Capuzzo-Sollum area, late April 1941: A motorcycle rider of the *Knabe* advance unit on a heavy BMW R 12 1935-41, 20 HP, 750 cc SV machine.

North Africa, Capuzzo-Sollum area, late April 1941: The *Knabe* advance unit at rest. In front are the heavy BMW R 12 1935-41, 20 HP, 750 cc SV cycles. Behind them are two captured English cars and a Kfz. 15. The soldiers are wearing the tropical version of the motorcycle coat, which is cut like the water-repellent coat but made of heavy khaki cotton.

North Africa, near Sollum, May 29, 1941: A British BMW motorcycle unit (captured machines) shortly before setting out on a new mission. The motorcycle rifleman in the sidecar is examining the German machine pistol.

North Africa, not far from Sollum, May 29, 1941: The leader of the British BMW motorcycle unit (captured machines) gives his last instructions.

North Africa, near Sollum, May 29, 1941: A British motorcycle unit on captured BMW R 12 machines. On the other side too, the heavy BMW was a very popular vehicle.

Libyan-Egyptian border, May 1941: An advance unit of the German *Afrika-Korps* rounds up prisoners. The motorcycle is a heavy BMW R 12 750 cc SV machine with sidecar.

The *Grossdeutschland* Division on the way to the eastern front. In front is a heavy BMW R 75 1940-44, 26 HP, 750 cc OHV cycle with driven sidecar.

Eastern front, Sunday afternoon, June 22, 1941: Friendly reception by the native population east of the San. The driver of this heavy BMW R 12 1935-41, 20 HP, 750 cc SV cycle belongs to a propaganda company of Panzer Group 1 (von Kleist).

Eastern front, late July 1941: During heavy fighting with the VIII. Mechanized Corps of General Riabyshev, southeast of Luck. An officer of the propaganda company of Panzer Group 1 (von Kleist) seeks cover from enemy planes; next to him is a heavy BMW R 12 1935-41, 20 HP, 750 cc SV cycle.

Eastern front, summer 1941: The advance unit of a regiment on a reconnaissance mission.

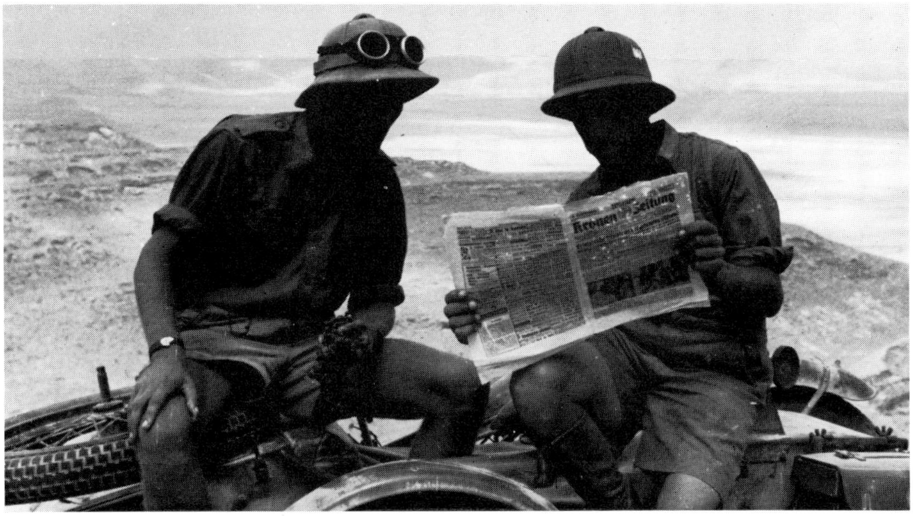
Kronen Zeitung

North Africa, summer 1941: "Successful fighting on the entire eastern front" say the headlines of the "Wiener Kronen-Zeitung." The two *Afrika-Korps* men have a black-white-red emblem on the right side of their tropic helmets, and the Wehrmacht eagle on the left. These tropic helmets were issued to all ranks as of the spring of 1941. They were uncomfortable, impractical and anything but suitable for motorcycle drivers.

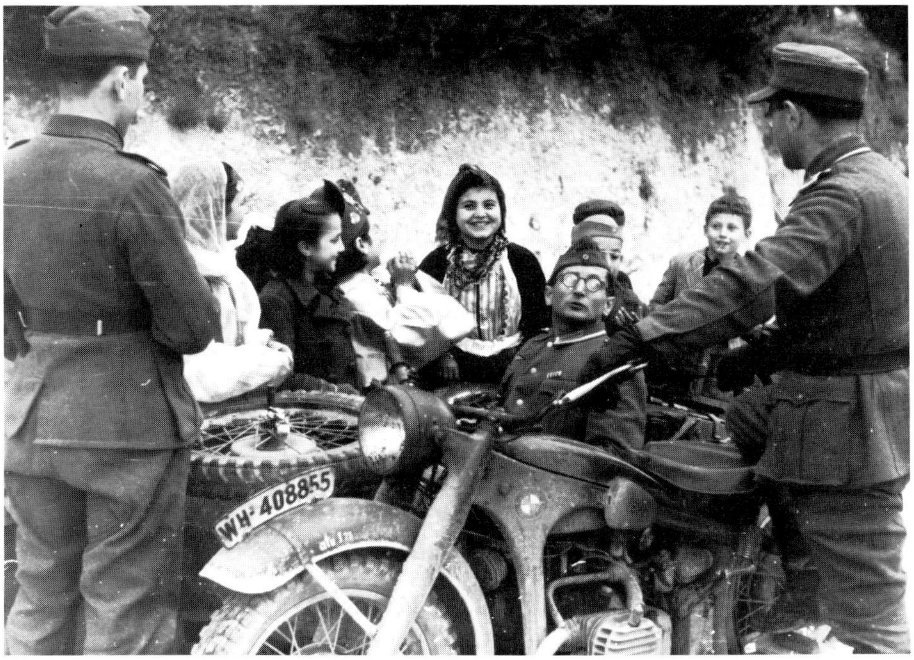
WH 408855

Balkans, summer 1941: A heavy BMW R 12 1935-41, 20 HP, 750 cc SV machine takes a break. The spare wheel is atop the nose of the sidecar.

Eastern front, summer 1941: A light motorcycle rifle troop with heavy BMW R 12 1935-41, 20 HP, 750 cc SV cycles on their way through a Ukranian village.

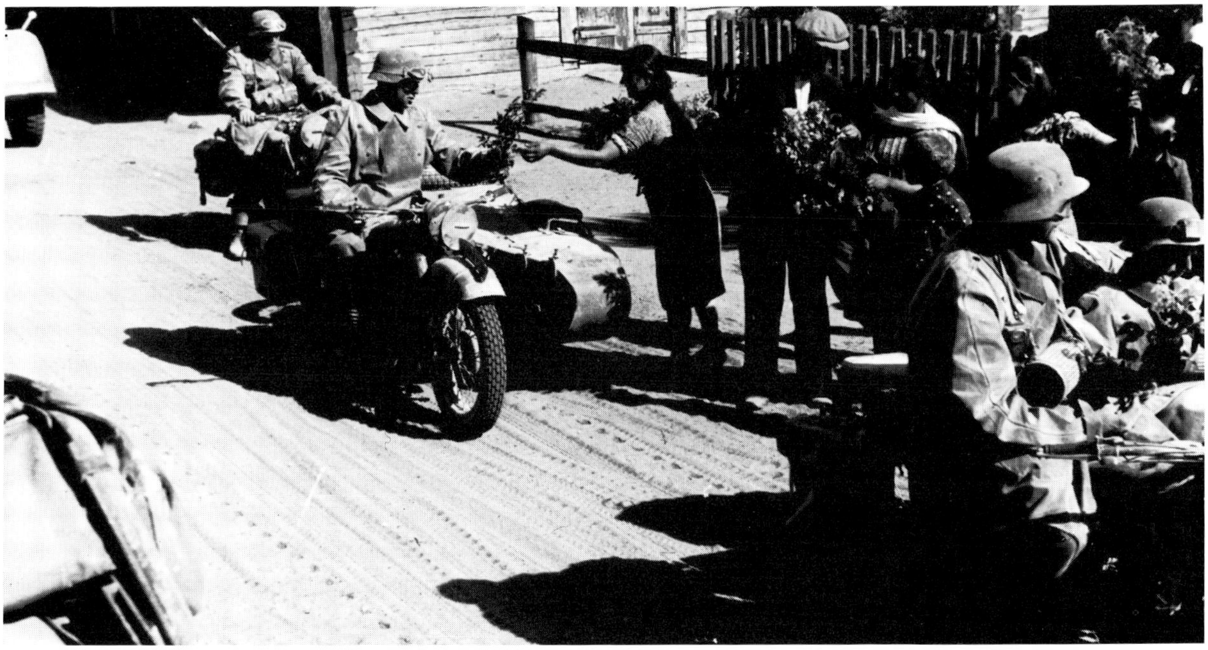

Eastern front, summer 1941: A motorcycle rifle company advances. In the background is a shot-down enemy truck column. At the left is a DKW 500 cc cycle with sidecar.

Eastern front, south of Smolensk, summer 1941: Two motorcycle riflemen bring a shot-down Russian air force lieutenant to the main dressing station on their heavy BMW machine.

Eastern front, July 1941: A heavy BMW machine of the 3rd SS Panzer Division "Totenkopf" on a road in the southern sector.

Eastern front, summer 1941: A regiment advances on the road. Al left is a messenger on a BMW R 4 1932-36, 12 HP, 400 cc OHV off-road sport Reichswehr cycle, at right a BMW R 12 1935-41, 20 HP, 750 cc SV machine.

Eastern front, summer 1941: The vehicle repair troop (company workshop) overhauls a heavy BMW R 12 1935-41, 20 HP, 750 cc SV cycle.

Greece, summer 1941: The motorcycle messenger takes a command from the company chief of a Panzer unit. In front is a medium BMW R 35 1937-40, 14 HP, 350 cc OHV cycle.

Eastern front, summer 1941: The heavy BMW R 12, 1935-41, 20 HP, 750 cc SV machine.

Eastern campaign, summer 1941: A motorcycle messenger on a BMW R 35 1938, 24 HP, 500 cc OHV cycle. At right is a 38 (t) tank, at left a Panzer II, Type A tank.

Eastern campaign, summer 1941: A motorcycle platoon discusses a mission. The company chief checks a map of the area. In front is a BMW R 12 1935-41, 20 HP, 750 cc SV machine.

Eastern front, summer 1941: The BMW R 12 1935-41, 20 HP, 750 cc SV cycle of an advance unit. A folded tent is strapped on the spare wheel of the sidecar, with a 20-liter fuel canister on top of it. At left in the ditch is a burning Soviet BT 5 tank.

Southern Bavaria, July 20, 1941: Testing a BMW R 75 in rough country.

Southern Bavaria, July 20, 1941: Testing the BMW R 75. "The machine runs superbly across country and has hitherto unknown potential to keep going in all situations."

Eastern front, summer 1941, somewhere in the southern sector: At left a heavy BMW R 12 1935-41, 20 HP, 750 cc SV machine, in the middle a messenger on a DKW NZ 350 cc cycle, at right an armored command car I, Type B.

Eastern front, summer 1941: The heavy BMW machine rolls ahead past a burning German Panzer II.

Eastern front, summer 1941: On the road with a heavy BMW machine. The group leader is riding in the sidecar.

Eastern front, summer 1941: An NCO of a police battalion with a heavy BMW R 12 1935-41, 20 HP, 750 cc SV machine, equipped with bumpers to protect the cylinders in rough country.

Eastern front, summer 1941: The BMW column of a motorcycle rifle company during the advance. Inside a village it encounters the enemy; the riflemen mount and prepare for combat.

Eastern front, summer 1941: A wounded Russian soldier is brought to a main dressing station by a German *Sanitäts-Feldwebel* (medical sergeant) on a heavy BMW machine.

Eastern front, 1941: A summer night on the road. A heavy BMW in a burning village.

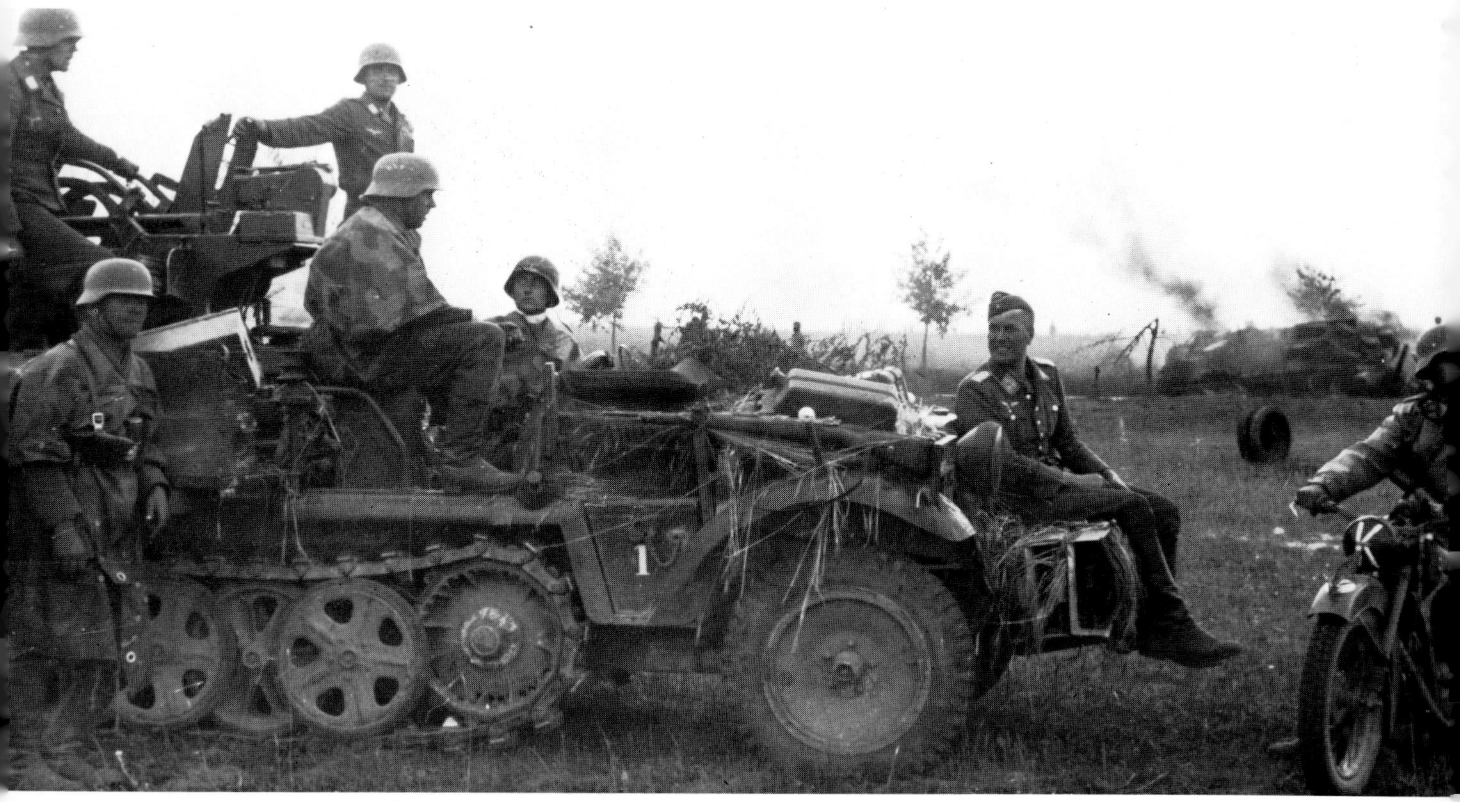

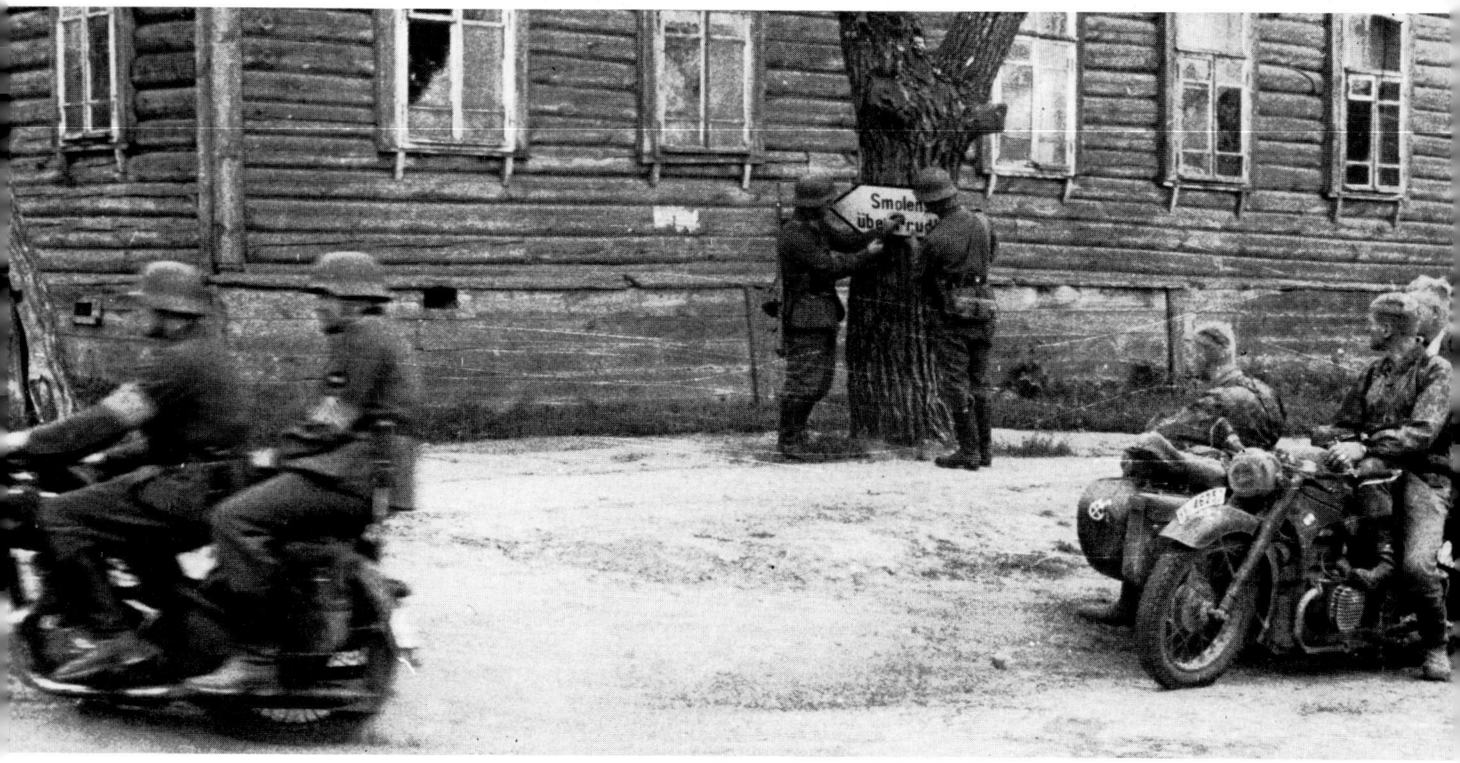
Smolen

« Eastern front, summer 1941: A motorcycle messenger is at right, on a BMW R 35 1937-40, 14 HP, 350 cc OHV medium cycle. In front is a self-propelled 20mm anti-aircraft gun pulled by a 1-ton towing tractor. In the background is a shot-up tank.

Eastern front, summer 1941: This *Gefreite* has really had "pig" (good luck). It unwillingly takes its last ride in the (driven) sidecar of a heavy BMW R 75 1940-44, 26 HP, 750 cc OHV cycle.

Eastern front, mid-September 1941: A military police patrol of the 3rd Panzer Division near Lochwitza has gotten its heavy BMW R 12 1935-41 20 HP, 750 cc SV cycle stuck. According to service instructions, "If the cycle has gotten stuck, dismount! Lift the front wheel out! Engage low gear and push!"

« Eastern front, summer 1941: The route marking service puts up a signboard. At right is a BMW R 12 1935-41 20 HP, 750 cc SV cycle of an SS motorcycle rifle company, at left a DKW NZ 350 cc machine.

Eastern front, September 1941: A motorcycle messenger of the 9th Panzer Division on a BMW near Mirgorod. The machine has a leather saddlebag on each side.

Eastern front, September 12, 1941: On this day the offensive advance of Panzer Group 1 (von Kleist) northward from the Krementshug bridgehead began. The soldiers of the propaganda company of Panzer Group 1, seen here pushing a heavy BMW R 12 1935-41, 20 HP, 750 cc SV machine, are getting a foretaste of what awaits them in Russia after the summer rain.

Eastern front, autumn 1941: Three exhausted soldiers of a medical unit get a well-earned rest in a heavy BMW R 12 1935-41, 20 HP, 750 cc SV cycle. The white G indicates the Guderian Panzer Group.

Naples, autumn 1941: The motorcycle riflemen of the German *Afrika-Korps* are greeted warmly. In front is a heavy BMW R 12 1935-41, 20 HP, 750 cc SV machine.

Eastern front, autumn 1941: The passenger shows the way. It's hard to advance on this road in a heavy BMW R 12 1935-41, 20 HP, 750 cc SV cycle.

Eastern front, autumn 1941: A heavy BMW R 12, 750 cc SV cycle with sidecar, of the supply unit of a Panzer division. The driver wears a motorcycle coat, the cycle carries a big leather case beside the rear seat.

Eastern front, 1941: After autumn rain it is not easy to get a machine out of the mud.

Eastern front, autumn 1941: Wet weather for a heavy BMW R 12 1935-41, 20 HP, 750 cc SV motorcycle.

Between Putivi and Orel, early October 1941: Motorcycle riflemen of an advance unit of the 17th Panzer Division, Panzer Group 2 (*Generaloberst* Guderian) in a trench. Heavy BMW R 12, 750 cc SV motorcycle with sidecar and machine gun.

Autumn 1941: Loading the heavy BMW R 12 1935-41, 20 HP, 750 cc SV cycle from a ramp onto a stake truck.

4

On a rainy day in the summer of 1941, a heavy motorcycle and sidecar pushed through the pathless hills of Upper Bavaria. A driver from the BMW Testing Department and his "fellow sufferer" in the sidecar tested the driving qualities of the newest version of the R 75. Here is the test report:

"Test drive with the R 75 motorcycle on July 20, 1941. The machine was driven 350 km, partly on Autobahn and roads and 25 km cross country. The sidecar was occupied. On normal roads, I experienced considerably better handling characteristics than our previous R 12 and R 71 models, especially in taking curves. In particular, taking curves affords considerably more safety and requires less power.

"After a distance of 150 km, though, seat difficulties appear, which are to be attributed to the fact that the saddle inclines in the direction of motion, and with a weight of 75 kg, this tendency was not overcome. Since with a normal load a suitable position of the saddle is not attained, the driver slides forward, which soon leads to unpleasant feelings of weariness. With wavy or bumpy road conditions it also occurs that the machine, at the prescribed air pressure, inclines to bounce, particularly in front.

"The shifting has not improved from that of the R 71, not to assert the opposite. According to my experience, the shift distances have also grown somewhat greater. Off the road, the foot shift lever can be used only for shifting into reverse, as when one shifts forward, one is in danger of getting one's foot stuck between the shift lever and the ground. The brakes are very good in their functioning and remain constant even after long use.

"The machine is excellent off the road and offers, in all situations, hitherto unknown potential for moving ahead. Especially with a stuck front wheel, and in connection with the differential lock, an unconditional solution to this awkward situation is provided by the reverse gear. The fuel consumption for 350 km, at an average speed of 55 km, is 8.5 liters per 100 km."

Here are some of the special design features of the BMW R 75:

The two-cylinder motor, with a displacement of 750 cc, has a square bore-to-stroke ratio, that of 78 x 78 mm. The "dropped" valves are activated by pushrods and rockers, all these parts being fully enclosed. On the forward end of the crankshaft the base of the generator is attached, while the magneto ignition is driven by diagonally toothed gears. To avoid long intake ducts, each cylinder has its own needle-jet carburetor. The fresh-air ducting leads to a newly developed felt-bellows air filter with a large surface, which is located in a low-dust air zone over the tank and can be shaken out and thus cleaned with two motions.

Despite its small dimensions, the gearbox has room for not only four forward gears but also reverse and overdrive through which off-road gears and an off-road reverse gear are added. It is this multitude of gears that makes the cycle suit all possible loads. The power is carried from the gearbox to the rear axle by a rubber driveshaft to a bevel gear and from there not directly to the rear wheel, but via a spur gear.

The manually locking spur gear differential divides the power in correct proportions between the rear wheel and the less heavily loaded sidecar by means of the size of the gears (and the resulting lever arms). From the differential, a torsion-bar shaft leads to the spur-gear drive of the sidecar wheel.

The R 75 has a rigid double-tube frame with a sheet-steel box backbone, so that individual parts can be changed easily. The smooth telescoping front fork with enclosed coil springs provides light weight for the unsprung masses. The sidecar wheel is made as a swinging wheel and sprung with a torsion bar that encloses the driveshaft. The steel body of the sidecar is mounted on leaf springs.

The hydraulic brake for the rear wheel and the sidecar wheel, new to motorcycle construction, can be cited as special equipment. It has the advantage of self-acting and safe equalization. The large 4.50-16 tires with off-road profile can be equipped with snow chains for rough country or muddy roads.

In addition, the new BMW R 75 has several important advantages. Three of them are especially worth knowing in detail: the sidecar drive, the felt-bag filter and the demountable frame. The sidecar drive:
Anyone who sits in the saddle of the vehicle for the first time will have become aware after a short time of the unpleasant inclination of his cycle to pull toward the sidecar side if the adjustment is not right. The cause: The asymmetry of the vehicle, that is, the center of gravity lies near the cycle, is displaced to the side. Yet because of its inertial opposition, the center of gravity creates an opposing force that tends to hold back the sidecar.

Thus when a forward impulse pushes on one side and the opposing power of the center of gravity works in the opposite direction, a turning motion is created by which the vehicle tries to turn toward the sidecar. A slight pressure to the right occurs, but it can be almost completely eliminated by the movement of the cycle (lateral outward tendency).

When the sidecar wheel is also driven by a rigid shaft, the vehicle runs straight ahead. Since both drive wheels are driven by a rigid shaft with the same power and at the same speed, and the sidecar wheel is farther away from the vehicle's center of gravity than the drive wheel, then it has the tendency to run ahead, which causes a pull to the left.

The installation of a spur-gear equalizing drive in the R 75 equalizes the different turning speeds, prevents the sidecar from pulling and makes the steering easier than that of vehicles with one-wheel drive. Locking the equalizing drive makes rigid-axle drive possible, in order to let the wheels take hold effectively on poor surfaces that encourage skidding.

The most important advantage of the off-road characteristics is provided by the two-wheel drive, since the driving power of the motor is divided between two wheels. Even with poor adhesion, the spinning of the wheels is prevented, and all difficulties occurring off the road, such as steep inclines, swamps, loose stones, sand and snow, can be overcome.

Even on uneven ground, the sidecar wheel

retains adhesion, which is made possible by a swinging arm that carries the sidecar wheel. Always linked with the driveshaft, the driven sidecar wheel swings up or down, whereby all the jolts of the uneven surface are absorbed without being transmitted to the sidecar frame.

The fine desert sand that abrades the motor, and the fine dust of the Russian roads, which can cause considerable abrasion to the cylinder and piston rings after even a short time, gave the designers of the vehicles the impetus to develop countermeasures. One of them is the felt-bag air filter. The damp-air filters formerly used in motorcycles, with a degree of 40-50% efficiency, are perfectly sufficient for German road conditions. But columns driving in Russia raise clouds of dust that are almost impenetrable; conditions that those in North Africa are scarcely worse then, since there it is possible to spread a column laterally. Even a moist-air filter followed by an oil bath, such as was formerly used, cannot offer sufficient protection against abrasion on account of the abrasive action of the sand and thus are only effective to the extent that at high engine speeds the intake air reaches a correspondingly high speed.

Better results can be obtained only with centrifugal oil-bath filters, as even the finest dust particles are pulled out by centrifugal force and come to precipitate through the oil pressure and spray as long as the oil is clean and fluid. The unpleasant feature of these oil filters: the necessity of having to carry oil and gasoline along to clean them. In heavy dust, the filter and oil get dirty very quickly and the filter loses its effectiveness. The driver must therefore often think of cleaning the filter in Russian road conditions.

These experiences led to the development of a felt-bellows filter that was installed previously only in larger vehicles but could not be installed in a motorcycle because of its large dimensions. It was possible to design such a felt-bellows filter for the R 75. Tests showed that the space over the fuel tank was most nearly free of dust and water.

The felt-bellows filter is mounted on the fuel tank with a sheet-metal cowling over it — similar in shape to a steel helmet — and so that this cowling does not rest completely on the tank. This creates intake openings for the air. The filter felt, because of its accordion form, has an unusually great surface so that, even when dirty from long driving distances, the average amount needed for air passage is still available. only when very high demands are made on the filter does the resistance increase: the passage becomes less, the gasoline-air mixture heavier. The fuel consumption increases, but the cylinder abrasion does not, because the cleanliness remains the same as before. The great advantage of the felt-bellows filter: At any air speed, thus at any engine speed, it is equally effective. Measurements carried out in unfavorable conditions on the eastern front gave surprising results: Practically the entire amount of dust was separated out.

Most important for the motorcycle driver in action at the front is the simple cleaning of the felt bellows without needing any tools, oils or gasoline for cleaning. The protective cowling is raised, the felt bellows taken out and simply shaken out. The dust on the outside of the felt falls off, and this takes care of the whole cleaning. The testing of this felt-bellows filter led to a further improvement which was installed in all BMW R 75 motorcycles: making starting easier at low temperatures. In normal use the inducted air passes through the felt bellows and then through openings in a tube that projects into the bellows and into the two intake ducts. This tube is now formed as a rotary pusher by means of which the inlet openings in the tube can be completely or partly closed by a lever that projects from the protective cowling. When the lever is placed at "start", the inlet openings close and the access of inducted air to the intake ducts is blocked. When the motor is started, the resulting low pressure pulls more fuel to the carburetor and provides a richer mixture for the starting process. This avoids the "jumping" of the carburetor to enrich the mixture.

Only when the motor fires and the low pressure begins to rise does a flutter valve high in the tube open to let somewhat more air in. When the motor is warmed up, the lever is set as "drive" and the induction openings in the tube are opened for full flow of cleaned intake air. In cold weather the intermediate setting of the lever allows a partial closing of the induction openings and the creation of a fuel mixture suited to the temperature conditions, whereby even running and high torque are attained even in a cold motor.

The third new design, based on war experiences, is the demountable frame of the BMW R 75 motorcycle.

Its great advantage: The possibility of easy replacement of any of the eight separate main components. If a part of the frame is damaged, the whole machine does not need to be dismantled as in the case of a one-piece frame; only the damaged part of the frame need be replaced. Once can also repair the frame quickly with a few tools, whereas a special workshop is needed to prepare a complete frame.

A further advantage: easier storage of parts. A demountable frame can be stored in a much smaller space than a complete frame unit, and a man can carry the dismantled frame under one arm with no trouble. Thus, for example, a motorized unit needs only three dismantled frames of the BMW cycle. From these parts they can repair 10, 12 or more cycles as long as the same piece must not be replaced in every repair job.

The frame of the BMW R 75 is composed of three triangular frames that are all braced against each other and can offer strong resistance. Since the connecting points are screwed and not welded, the braces remain free of bending.

To meet the high pressure on the upper frame member, it is formed as a box member and shows a very high profile in sheet metal. At the rear end of the box member, the profile attains a nearly round cross-section that is best suited to absorb twisting. At this end a cross member is welded on that transmits any twisting pressure through lever action in the form of pushing and pulling pressure to the saddle brace and the gearbox mount.

Other forces affect the hub of the rear wheel, partly through forward motion and braking. These pressures are also transmitted to the braces, and specifically those of the rear triangle, and are changed into pushing and pulling pressures that are transmitted on to the adjacent triangles.

Eastern front, November 1941: The heavy BMW machine in a traffic jam on the road.

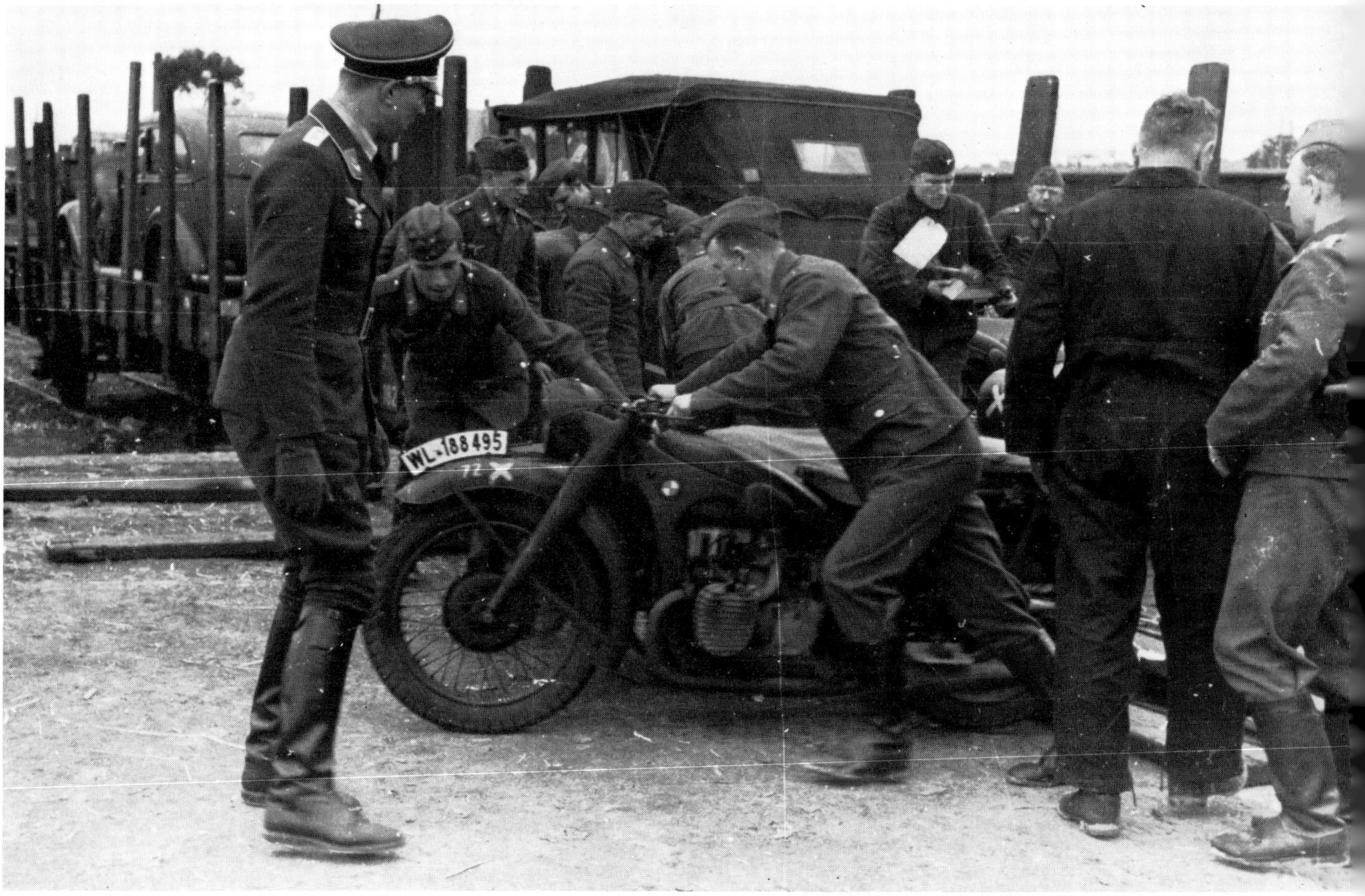
WL-188495

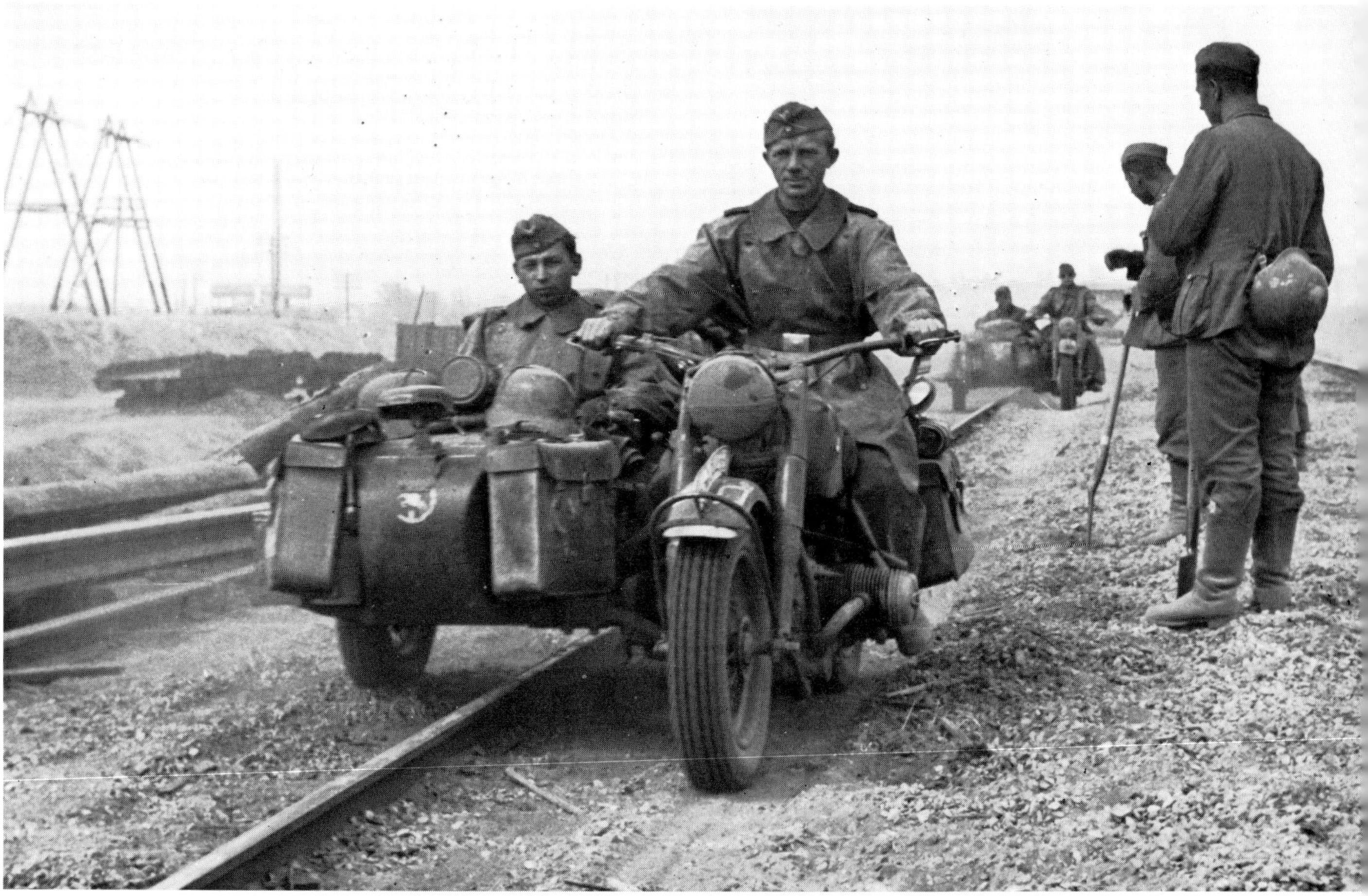

« Eastern front, central sector, autumn 1941: The heavy R 12 1935-41, 20 HP, 750 cc SV cycle of a Luftwaffe unit is unloaded.

« Eastern front, autumn 1942: The advance unit of the 24th Panzer Division on the march along the railroad tracks with heavy BMW R 75 1940-44, 26 HP, 750 cc OHV machines with (driven) sidecars. On the front of the sidecar hull between the packs is a camouflaged headlight, not customary on motorcycles.

North Africa, the deserted positions near El Alamein, November 5, 1941: That was once a speedy BMW R 12. The motorcycle was flattened by a British tank.

North Africa, December 15, 1941: Into action on a heavy BMW captured from the German *Afrika-Korps* goes this British motorcycle patrol in the vicinity of Tobruk.

North Africa, Tobruk area, December 3, 1941: A reconnaissance unit of the German *Afrika-Korps* is captured by the soldiers of a New Zealand unit. In front is the heavy R 12 solo cycle, behind it a BMW R 12 with sidecar and a heavy personnel carrier.

Smolensk, winter 1941-42: A motorcycle group on the march. From left to right, a BMW R 71 1938, 22 HP, 750 cc SV, a Zündapp KS 600 1939, 26 HP, 600 cc OHV, and a BMW R 12 1935-41, 20 HP, 750 cc SV.

Eastern front, winter 1941-42: The field workshop of a vehicle repair company. In front is a heavy motorcycle with (driven) sidecar, BMW R 75 1940-44, 26 HP, 750 cc OHV.

Eastern front, winter 1941-42: Advance of a heavy motorcycle rifle unit over snowy country. In the foreground is a BMW R 75 1940-44, 26 HP, 750 cc OHV cycle with (driven) sidecar.

El Agheila, Libya, December 1941: A motorcycle driver of the German *Afrika-Korps* (reconnaissance unit) at his morning ablutions on a heavy BMW R 12 750 cc SV cycle with sidecar.

Benghazi area, December 1941: A short siesta for an anti-aircraft unit of the German *Afrika-Korps*. In front is a heavy BMW R 12 750 cc SV cycle with sidecar.

Tripoli, winter 1941-42: A flat tire draws a crowd. The rear wheel of a heavy BMW R 12 1935-41, 20 HP, 750 cc SV cycle is being pumped up. The driver wears the tropical uniform, typical of the *Afrika-Korps*, made of light denim, and laced canvas boots.

Germany, Christmas 1941: Some cheery mail has reached a Luftwaffe unit. Heavy BMW R 12 750 cc SV cycle with sidecar.

Eastern front, winter 1941-42: A heavy Zündapp KS 600 cycle and sidecar with improvised chalk winter camouflage at a division command post. The passenger examines the map in a folding map case. The three soldiers wear improvised snow coats, typical of the first Russian winter. In the background is a medium personnel carrier (Kfz. 15).

Eastern front, spring 1942: The tents were set up as cover from an unexpected snowfall. At left is a heavy BMW R 75 1940-41, 26 HP, 750 cc OHV cycle with (driven) sidecar; on the sidecar is the tripod mount for a heavy machine gun.

Eastern front, spring 1942. Men of a Panzer regiment have turned out for a company roll call before their regimental commander. In front is a heavy BMW R 12 1935-41, 20 HP, 750 cc SV cycle with a G-LT special device on the sidecar. In the background is a Panzer IV tank (75mm KwK L/24 tank gun), Type D.

Eastern front, spring 1942: A messenger on a medium BMW R 35 1938, 14 HP, 350 cc OHV cycle. When fording water, the instructions said, "Select low gear! Drive slowly! Don't choke the motor!" The driver wears a water-repellent gray-green rubber coat. Its tails can be wrapped around his legs from the rear and then be buttoned so the driver can move better on his cycle and his coattails won't get tangled in the spokes. The collar is of field-gray cloth, and on the front of the coat are two pockets protected by particularly large flaps.

Eastern front, spring 1942: Everybody helps out, even an officer (second from left). A heavy BMW R 12 1935-41, 20 HP, 750 cc SV cycle.

Eastern front, spring 1942: A motorcycle messenger on a heavy BMW R 12 1935-41, 20 HP, 750 cc SV cycle shifts into gear. His equipment includes a water-repellent motorcycle coat, 98k carbine and cloth gloves.

Eastern front, spring 1942: Thanks to the sidecar drive, the driver of this heavy BMW R 75 1940-44, 26 HP, 750 cc OHV cycle with sidecar has the chance to get his machine out of the mud himself. This soldier, from a tank unit, wears a light army camouflage jacket and black 1942 model field cap.

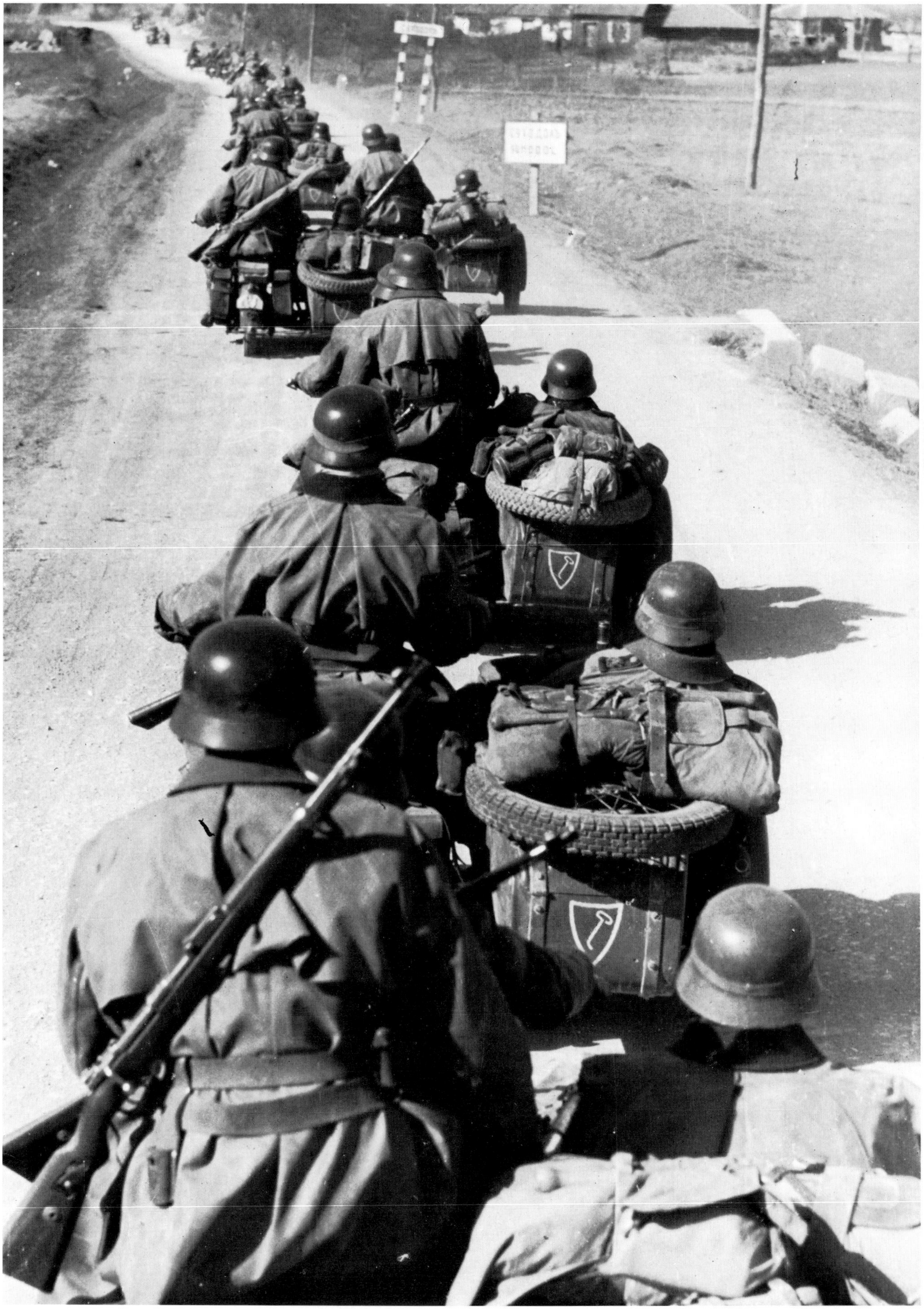

Eastern front, 1942: A heavy BMW R 12 1935-41, 20 HP, 750 cc SV machine, with a Zündapp in the background. Motorcycle riflemen of a paratroop unit at rest, wearing steel helmets with camouflage covers and paratrooper jackets.

« Eastern front, spring 1942: A motorcycle battalion of the 1st SS Panzer Division, *Leibstandarte Adolf Hitler*, on the march with their heavy machines, 1935 model steel helmets with rubber bands to hold camouflage material, motorcycle coats, 98k carbines, spare wheels, pack pouches with combat packs and trenching tools among their equipment.

Balkans, spring 1942: A motorcycle messenger with mail sack mounts his BMW R 12 1935-41, 20 HP, 750 cc SV cycle. The tails of his water-repellent coat have been wrapped around his legs from behind and buttoned in place.

Demonstrating the handling qualities of a heavy BMW R 12 1935-41, 20 HP, 750 cc SV machine by a unit of the SS Mountain (Gebirgs) Jäger Regiment of the 23rd SS Mountain Division.

Paris, spring 1942: The motorcycle driver of a Luftwaffe unit with a heavy BMW R 12 1935-41, 20 HP, 750 cc SV machine. The headlight has been painted instead of being masked with a cloth cover.

North Africa, spring 1942: A short rest on a desert road. In front are two heavy BMW R 12 1935-41, 20 HP, 750 cc SV machines, at right a medium personnel carrier.

Yugoslavia, spring 1942: After an accident a medium BMW R 35 1937-40, 14 HP OHV cycle is salvaged. A leather pouch is attached to the front fender.

Eastern front, spring 1942: An army staff's map office. The courier is just getting the map container which he is to deliver. In front is a heavy BMW R 12 1935-41, 20 HP, 750 cc SV cycle.

Eastern front, spring 1942: The courier on the way from a map office to a unit on his heavy BMW R 12 1935-41, 20 HP, 750 cc SV cycle. On his back beside his 98k carbine is the map container of sheet aluminum.

Eastern front, northern sector, road conditions on May 30, 1942. A motorcycle messenger gets information about his march route.

Eastern front, summer 1942: Fording a stream in the central front sector. The engine of this BMW R 12 1935-41, 20 HP, 750 cc SV has quit mid-stream.

Eastern front, summer 1942: A wounded man is being taken to a dressing station on a heavy BMW R 75 1940-44, 26 HP, 750 cc OHV cycle with (driven) sidecar.

Eastern front, southern sector, summer 1942: The (driven) sidecar of a heavy BMW R 75 26 HP cycle, with bumper and pack bracket.

The crew of a heavy BMW R 75 1940-44, 26 HP, 750 cc OHV cycle with sidecar repairing the brakes of the sidecar. These paratroopers wear steel helmets with covers and light camouflage jackets.

The vehicle workshop troop of a paratroop unit repairing the sidecar of a heavy BMW R 75. On the front of the sidecar is the bracket for leather pack pouches. The seat has its cover on. At the upper right is a medium personnel carrier.

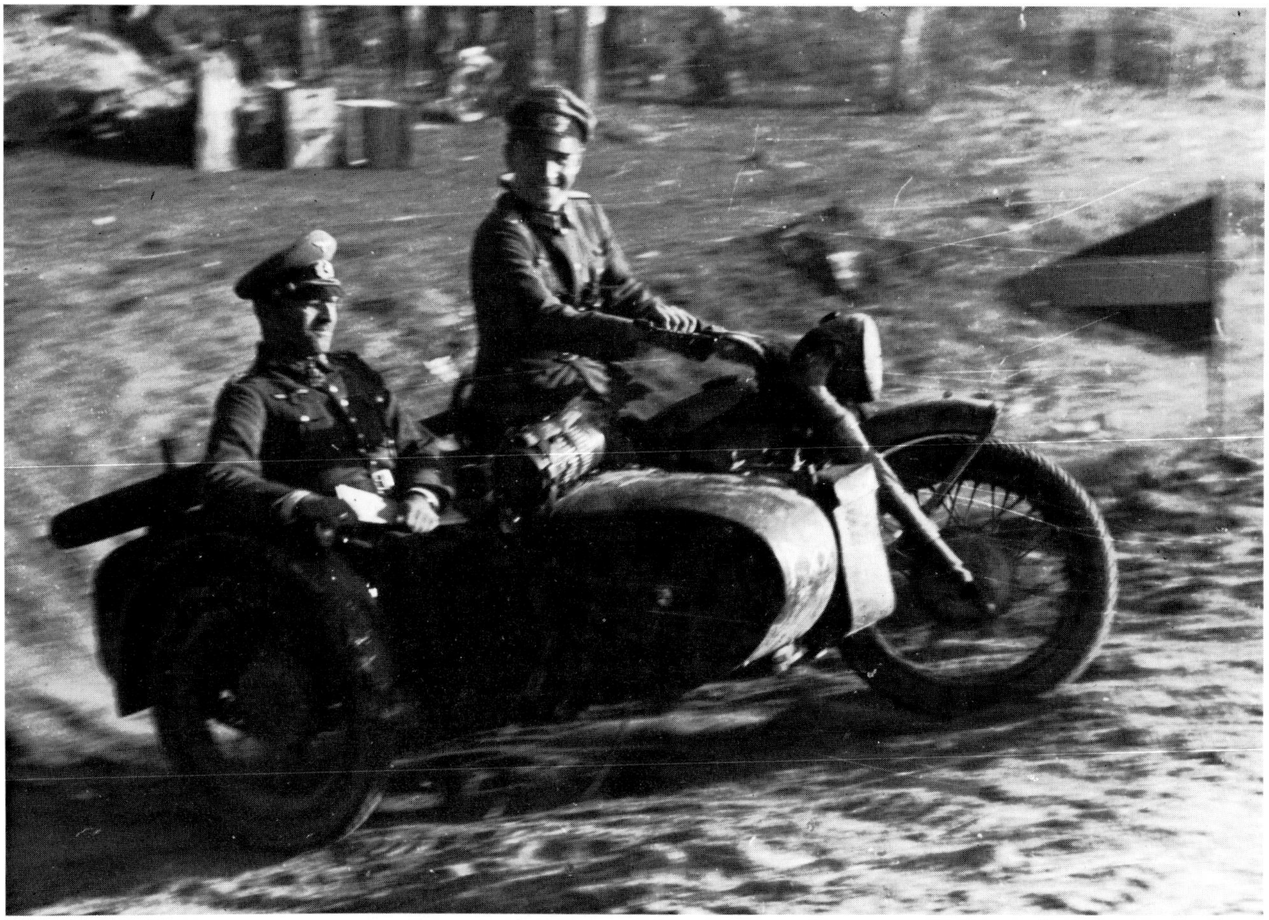

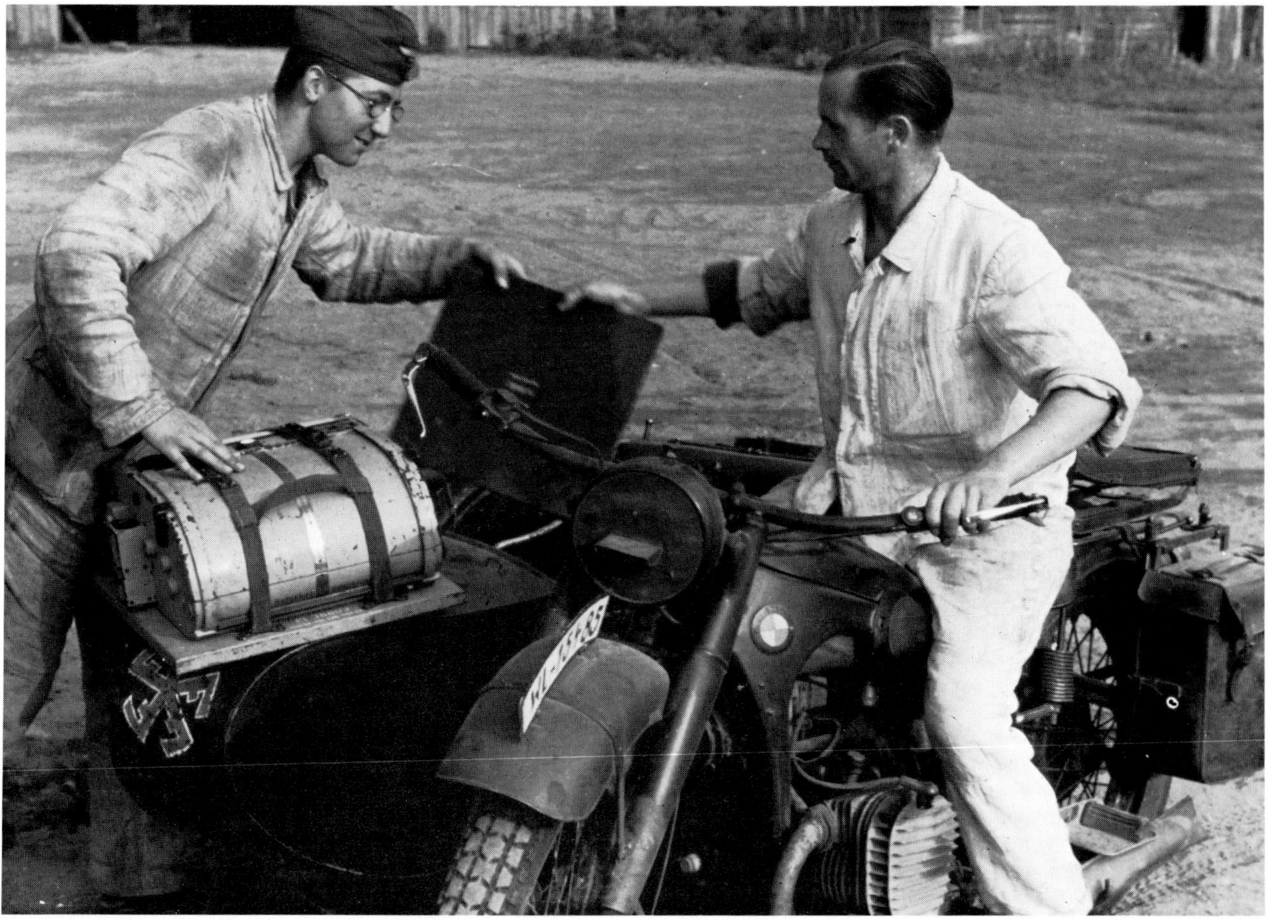

Opposite page, above:
Eastern front, summer 1942: A major general in the sidecar of a heavy BMW R 12 1935-41, 20 HP, 750 cc SV cycle on his way to the command post of an infantry division.

Opposite page, below:
Eastern front, summer 1942: A motorcycle driver of the 5th Echelon of Luftwaffe Reconnaissance Group 122, on a heavy BMW R 12 750 cc SV machine, with a camera for air photos on the sidecar.

Eastern front, summer 1942: The driver is always responsible for maintenance. Here he has a defective tire on his heavy BMW R 12 1935-41, 20 HP, 750 cc SV cycle.

Eastern front, summer 1942: A motorcycle rifle battalion makes its way forward laboriously on a dusty road.

Eastern front, summer 1942: Two heavy BMW R 12 1935-41, 20 HP, 750 cc SV machines meet on a narrow mountain road. The driver of the machine at right has covered his 98k carbine with a canvas dust cover. The cycle in front belongs to the II. Rifle Company of the 6th Panzer Division.

Greece, summer 1942: Two navy motorcycle drivers with a heavy BMW R 12 1935-41, 20 HP, 750 cc SV cycle.

Russia, summer 1942: A heavy BMW R 12 1935-41, 20 HP, 750 cc SV cycle, with mail and reading matter ("Der Kurier Ost") stowed in its sidecar.

5

The heavy BMW R 75 machine had one powerful competitor, despite all its advantages: the Zündapp KS 750. This also had sidecar drive and reverse gear. In the years from 1940 to 1944, a total of 18,635 of them were built — almost 2000 more than the BMW R 75. The BMW and Zündapp firms, whose total production met the war needs of the Third Reich, made a private agreement — a sort of armistice for the time being — in their battle for foreign markets:

March 31, 1941

Confidential!
VK 740 Dr. E./Ot.
895, 868, 720, 760 and 516
Du. an P 542, 742, 741, 615
Exporting the extra-heavy motorcycles
(BMW R 75)
We have made an agreement with the Zündapp Works regarding the future exports of both firms of extra-heavy production motorcycles. Enclosed you will find the text of this agreement, in which the Zündapp firm has assured us:
1. "Until October 1, 1941 any advertising measures and sales negotiations with interested foreign parties regarding future delivery of the extra-heavy 750 cc motorcycles will be avoided both by you (BMW) and by us (Zündapp)."
2. "Should a change in the political situation or an expiration of the present official terms make a breaking of this agreement necessary before then, then we will notify you, and you us, accordingly."

We ask for friendly consideration and acceptance of this agreement on the part of your department.
VK 740 (from an internal BMW factory memorandum).

After the Bayerische Motorenwerke had disposed of their export troubles in this way, they dedicated themselves all the more to their most important customer at the time: the Wehrmacht.

Since the war had caused limitations on civilian use of motor vehicles, the BMW customer service organization had been devoted to fulfilling Wehrmacht requirements. Through the constant participation of the firm's employees in Wehrmacht field training and the motor sport activities of the National Socialist Motor Vehicle Corps (NKSS) and the German Automobile Club (DDAC), valuable experience could be acquired for military use.

The main tasks of the BMW customer service department since September of 1939 were now the training and supplying of the motorized troops, establishment and maintenance of the repair shops and spare-parts supplies, so that the BMW customer service department did not limit itself to action on

German territory, but spread out to just behind the line of battle. The knowledge of the most important functions and the maintenance of motor vehicles, like that of any other equipment in use at the front, is a prerequisite of lasting readiness for use. Thus the greatest emphasis is placed on the training of repair and driving personnel in the Wehrmacht units, and a systematic training of drivers and workshop personnel was carried out.

In factory training at BMW, the technical personnel of motorized units and the instructional personnel of the Wehrmacht were familiarized with the qualities of the BMW products. For special training, cars with trailers were made available, likewise equipped with instructional material, special tools, blackboards and cutaway models.

The maintenance of the motorized parts of the Wehrmacht was kept up until just before the front, where the modern workshop trucks offered great support. Through them, not only could an assortment of special tools be carried, but also machines and equipment that guaranteed conscientious maintenance of the vehicles. A further task of these "mobile workshops" was the practical training of the personnel and the development of BMW specialists. The training of motorized troop units of the allies, such as Rumania and Hungary, also required a major part of this action.

A further support of the fighting troops consisted of the repair shops in the occupied territories; thus long trips of specialist personnel and spare parts for BMW vehicles could be avoided. And the engineers could convince themselves of the demands on vehicles in war service. The very determination of the frequency of disturbance in the readiness of spare parts is extremely important.

The expedient handling of spare-part distribution, often with a daily involvement of 800 and more positions, is the work of these BMW factory mechanics in service near the front, who in addition to practical work also train the driving and workshop personnel in terms of economical consumption of materials.

But even the most careful customer service and maintenance, when there was time for it between actions at the front, and as regular inspections as possible, were not able to change the fact that the vehicles on the eastern front were stressed in a hitherto unimaginable manner. To keep this damage within bounds, BMW gave out a brief guide to motor vehicle examination.

BMW R 75
Brief Guide to Vehicle Examination

Front fork, Wheels:
1. Check the fork bearing in the steering head. Procedure: Set cycle on rear standard and weight the luggage carrier (with the driver) until the front wheel hangs free. Completely loosen steering damper. While checking (by lifting the fork by the front fender), the bearing must be completely lacking in play but fall freely to either side. Eliminate play by loosening the wing bolt from the steering damper, loosening the cap bolt (use a normal 32 mm SW box wrench) and adjusting the flat adjusting screw between the steering plate and frame (special wrench on the vehicle). Tighten cap bolt very securely after finishing adjustment.
2. Check steering damper for trouble-free functioning, repair if necessary. Procedure: Front wheel must hang free (see Point 1), pull steering damper, swing fork to both sides.

There must be no play at the contact points on both sides of the steering damper. Eliminate any play by welding or bending the contact points.

3. Check oil filling of the telescopic fork.
Procedure:
a. R 12 fork: remove control screws, place cycle on both wheels, insert dipstick and check oil level. Oil depth 30 mm. Oil quantity per fork half 120 cc of shock-absorber oil.
b. R 35 fork: by pressing the rubber sealing sleeves down on the fork tubes it can be determined easily whether there is oil in the fork. Oil quantity per fork half 150-170 cc motor oil.
c. R 51-R 75 fork: unscrew closing screws on the steering plate (36 mm box wrench) and move the guide linkage up and down. If a "pumping" occurs, the oil is sufficient. In case of doubt, drain the oil and refill. Oil quantity 80-100 cc winter motor oil!

4. Check spokes, tighten if necessary.
Procedure: beginning at the valve, press the pairs of spokes together to see if spokes are loose. These should be tightened — if necessary, with an adjustable wrench.

5. Lubricate wheel hub, at least every 5000 km.
Procedure: remove wheels, press in natron-lathered oil such as Gargoyle-Mobil Compound No. 5 or Military Uniform Oil, but no more than 10 to 12 strokes, as the hub will be overfilled otherwise. If oil comes out the sides during lubrication, then the hub is strongly overfilled and the brake will be smeared.

Motor, Gearbox, Driveshaft
1. Add motor oil and change it at least every 2000 km (in winter every 1500 km), for off-road use every 1000 km (frequent oil changes prevent premature wear).
2. Check gearbox oil. Oil level up to the bottom threading of the filler opening. Fill with motor oil.
3. Check driveshaft oil. oil level up to the bottom threading of the filler opening. Fill with Mobil oil EPWI; if not available, use heavy motor oil.
4. Check screws and bolts on the motor, especially motor attachment screws, cylinders, cylinder covers and heads, carburetor intakes.
5. Check valve play. Minimum play (for Wehrmacht): intake 0.15, exhaust 0.15 for all models. (Some valve noise can be accepted, like burned-through valves).
Procedure: Turn motor with crank until the piston of the cylinder to be adjusted is at height of compression stroke. Check play with leaf tester, adjust if necessary.
6. Clean float housing, jets and air filter, dip air filter in oil before installation.
7. Check moving parts, adjust and oil. Play of clutch Bowden cable approximately 5 mm, on the others approximately 1-2 mm.

Chassis, sidecar attachments, brakes:
1. Tighten all screws and bolts on the chassis, especially protective panel fastenings, saddle, fuel tank, luggage rack, front and rear axle, exhaust system.
2. Tighten sidecar attachments, screws and bolts on the sidecar hull, panels and frame.
3. Test brakes and adjust for trouble-free operation.
Procedure: Drive the cycle on a dry road with as good adhesion as possible at a speed of approximately 25 kph and test the front and rear brakes separately. With normal brake use, a good effect should be noticeable.

Braking distance to a stop should be no more than 16 meters for the front brake and no more than 10 meters for the rear brake.

On all cycles the brake adjustment is corrected by the brake adjusting screw on the Bowden cable of the front-wheel brake.

Electric System

1. Lubricate the generator at the lubrication points. Add motor oil at the lubrication point on one side until the oil level becomes visible on the other side.
2. Test the electric system and wiring, repair worn spots in the wiring with electric tape.
3. Test the battery, attach the contacts firmly and grease them.
4. If the battery is dead or defective, the motor can also be made to start with battery ignition by pushing.
Procedure: Attach the battery connector cable to the generator with clamp 51/30. Likewise attach to clamp 61; firmly attach cable 51 to connector clamp 61. Engage second or third gear and push the cycle.
5. Check spark plugs, correct spark gaps (gap for magneto ignition 0.4-0.5 mm, for battery ignition 0.7-0.8 mm.
6. Check distributor gap, correct if necessary. Normal gap 0.4 mm at full distributor stroke.

Despite regular service and inspection, various defects were bound to appear in the R 75. And after the Army High Command first issued a thoroughgoing report in the spring of 1942 on the faults in the various models found by the troops, BMW took measures immediately.

The faults found in the R 75 were eliminated by appropriate changes after they were made known. In machines already delivered, the BMW factory replaced the parts with improved versions. Units in which the exchange of these parts could not be made at the factory were equipped with sufficient quantities of the improved parts, along with instructions.

Here are some of the most commonly occurring faults found during the eastern campaign and the ways to eliminate them:

Fork coverings that come apart as a result of serious dent holes and damage the springs:

The lower, inner fork covering was lengthened in January of 1942 to the extent that the two fork coverings cannot come apart under any circumstances.

Loosening of the anchor on the anchor shaft of the foot shift:

The anchor lever was welded to the anchor lever shaft on one side. It happened that the level came loose, so that it was no longer possible to engage the gears with the foot shift, although the hand shift worked perfectly. Machines in which such a fault occurred had to be fitted with an improved anchor lever, a changed gear segment and a strengthened holding spring. This innovation was applied to production machines as of chassis numbers 753,350 (holding spring) and 755,040 (anchor lever).

Breaking of teeth on the starter bevel gear:

The box pressed into the gearbox cover on which the bevel gear of the starter shaft runs can be displaced by the axial pressure when the motor is started. This makes the bevel gear on the starter shaft get out of line with the bevel gear of the starter, which leads to damage of the teeth of both gears. To avoid this, a starter disc was placed between the gearbox cover the bevel gear of the starter shaft to transmit the pressure of the bevel gear directly to the gearbox cover.

The gearbox cover had to be reworked to have an exact running surface for the added disc. This starter disc can be made in thicknesses of 3.8, 4.0 and 4.2 mm. The starter disc has to be strong enough so that the box and the starter disc are linked. In the future, the gearbox cover would be delivered in a form in which the box did not stick out; then the bevel gear runs directly against the gearbox cover and the starter disc is not needed.

Breakage of the differential housing:

When the differential was locked while driving, without declutching and letting up on the accelerator, it could happen that the locking hook broke and thus the differential housing was cracked. In August of 1941 the hook was strengthened. In order to avoid an unintentional activation of the lock, it was provided with a bolt as of machine no. 755,901.

After the very first months of the eastern campaign it was clear that heat, dusty and muddy roads, and then frost, snow and, above all, the Russian soldiers, who attacked more and more determinedly in the winter of 1941-42, negated the fighting tactics of the motorcycle riflemen and required rethinking as to the use of the fast units.

The halftrack vehicles and VW personnel cars (Kübelwagen) appeared in steadily growing numbers, for despite victories, the motorcycle riflemen, even with their heavy machines on which they were exposed to the cold without protection, and in times of mud, could not keep up with four-wheel-drive and tracked vehicles. Then too, the enemy's type of combat could not be compared with the 1939-40 campaigns, and to offer opposition decisively, one must have sufficient firepower.

Thus only a few of the 15th (Motorcycle Rifle) Company, as well as the 17th Company, lasted through the winter of 1941-42. They were disbanded. In place of them, the regiments and battalions gained motorcycle rifle platoons that carried out reconnaissance duties, route planning and the like. To be sure, after 1942 the reconnaissance units of the motorized infantry divisions existed with their old structure and equipment, but the decimated motorcycle rifle battalions of the Panzer divisions were merged with them.

Undisturbed by these changes to the war's end, though, were the motorcycle messengers, those brave soldiers on whom the army could always depend in all weather and at any hour. Thus in 1943 individual, independent motorcycle rifle battalions still existed, but they disappeared gradually. In the same year the reconnaissance units of the Panzer divisions had been changed almost completely. They now consisted of one armored scout car company, three rifle truck companies, and one heavy company with antitank, infantry gun, engineer and cannon platoons. A unit so equipped, with its approximately 1200 men, 140 armored vehicles, 160 other trucks, cars and motorcycles, is a very capable combat unit that can handle the reconnaissance and the combat assignments within the framework of a Panzer division but also operate completely independently on its own. Organized and equipped in this form, the armored reconnaissance units fought on until the end of the war.

Germany, summer 1942: Motorcycles of an army propaganda unit. The two cycles are heavy BMW R 12 750 cc SV machines with sidecars.

North Africa, summer 1942: An *Afrika-Korps* motorcycle driver with his heavy BMW R 12 1935-41, 20 HP, 750 cc SV cycle. In the background is a wooden water tower.

Eastern front, summer 1942: The motorcycle riflemen of the 18th Panzer-Grenadier Division ferrying their heavy BMW R 12 1935-41, 20 HP, 750 cc SV cycle across a river in a rubber raft.

A motorcycle rifleman of the 18th Panzer-Grenadier Division rolls his heavy BMW R 12 1935-41, 20 HP, 750 cc SV cycle off the rubber raft on a wooden ramp.

Eastern front, late summer 1942: A break in the march toward Stalingrad. The heavy cycle with (driven) sidecar is a BMW R 75 1940-44, 26 HP, 750 cc OHV type belonging to the 24th Panzer Division, formerly the 1st Cavalry Division. At the right rear is a light Daimler L 1500 A truck with troop-carrier body.

Lemberg, autumn 1942: Members of a police unit, heavy BMW R 12 750 cc SV cycle with sidecar, and no pad on the passenger seat.

Opposite page, above: Bulgaria, summer 1942: The ration carriers of Panzerjäger Platoon 2, Panzerjäger Unit 560, wearing motorcycle coats, with thermos containers and their BMW R 12 1935-41, 20 HP, 750 cc SV machine.

Opposite page, below: The two ration carriers on the way to their unit.

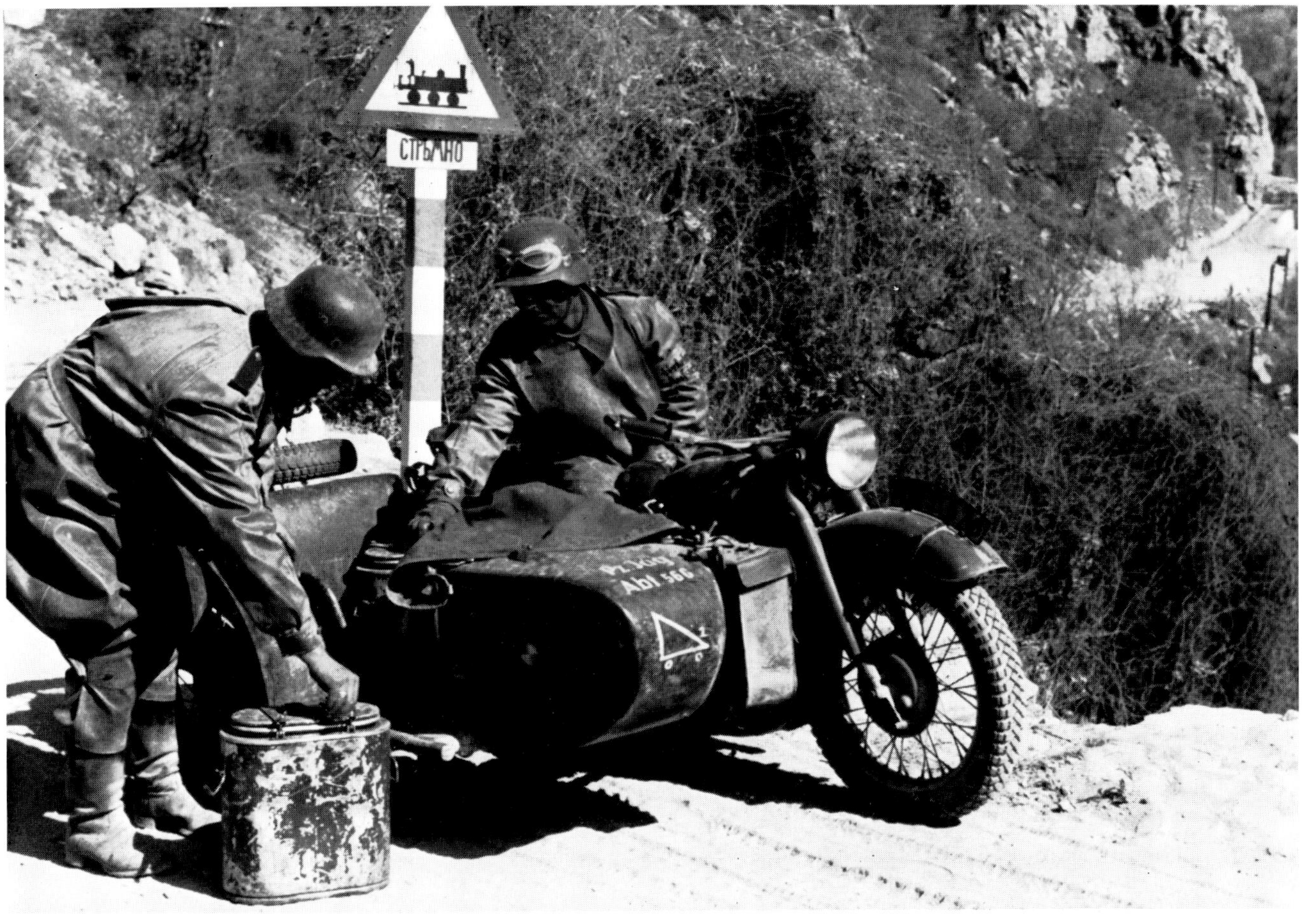
Abt 566

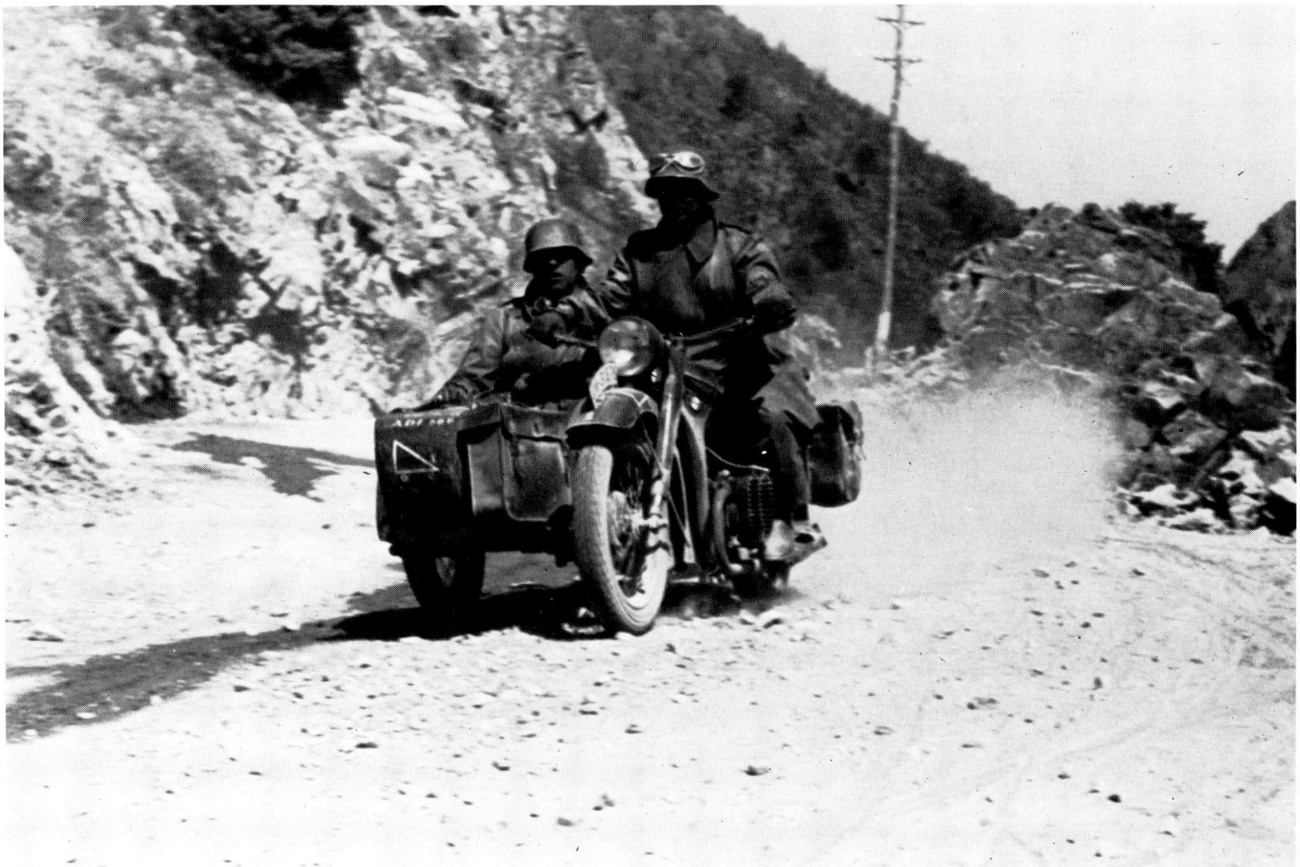

Paris, summer 1942: The regimental repair shop for motorcycles. In the foreground and at right are heavy BMW R 5 1936-37, 24 HP, 500 cc OHV cycles. In the left foreground are an FN 1000 cc Type M 12 SM and a Zündapp KS 600. On an assembly block in the background is an NSU.

Eastern front, summer 1942, near Vilna: A military police patrol has just put up a sign warning of danger from partisans. In front is a BMW R 12 1935-41, 20 HP, 750 cc SV cycle with a canvas saddlebag beside the passenger seat.

Sofia, summer 1942: A Hitler Youth member shows the way to the driver of a BMW R 12. The driver wears a 1935 model steel helmet, Leitz aviation goggles and a water-repellent motorcycle coat. His leather despatch case is under his right arm. A 20-liter fuel can is in front next to the sidecar.

Eastern Front, summer 1942: A motorcycle police patrol reads a map. On the fender is the sidecar headlight, at right the canvas sidecar cover, under the map a machine pistol.

Eastern front, late summer 1942: A unit of the 24th Panzer Division about to begin a march. The vehicles are camouflaged to avoid being seen from the air. In front is a motorcycle rifle unit on heavy BMW R 75 1940-44, 26 HP, 750 cc OHV cycles with (driven) sidecars. Beside them are a halftrack towing tractor (Sd.Kfz. 11) and a 3-ton Opel Blitz "S" truck. Beside the road are several Russian GAZ trucks. The second cycle from the front is a BMrd the Volga and Stalingrad.

Eastern front, somewhere east of Voronesh, late summer 1942: The 24th Panzer Division on the march toward Stalingrad. In front is a heavy BMW R 75 1940-44, 26 HP, 750 cc OHV machine, in the background a medium 8-ton towing tractor (Sd.Kfz. 7) with quadruple 20mm anti-aircraft guns, plus DKW motorcycles.

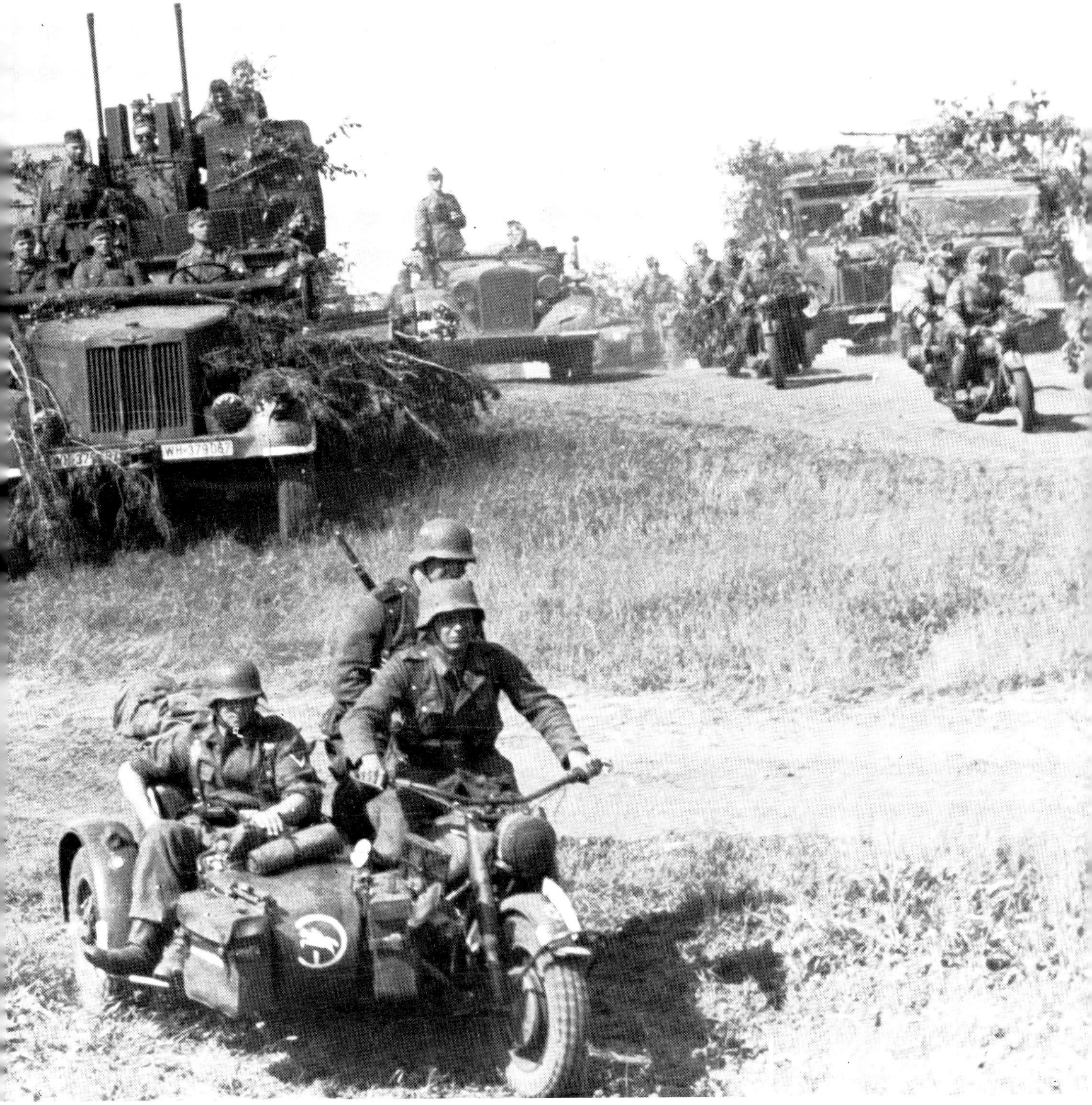
WH-379067

Eastern front, east of Voronesh, late summer 1942: The combat leader of the advance unit of a motorcycle rifle platoon of the 24th Panzer Division on a heavy BMW R 75 1940-44, 26 HP, 750 cc OHV cycle with (driven) sidecar charts his course to his combat position on the map. In front, on the driver's right, is the accelerator handle with the hand brake lever. On both sides of the sidecar are leather saddlebags. "The Jumping Rider" is the emblem of the 24th Panzer Division, formerly 1st Cavalry Division.

Eastern front, somewhere east of Voronesh, late summer 1942: The combat leader of a motorcycle rifle platoon of the 24th Panzer Division observes the enemy position. In the background are two BMW R 75 1940-44, 26 HP, 750 cc OHV cycles. The leader carries his sidearm very irregularly, over his right shoulder on a pack strap, and his magazine pouch hooked on in front. His MP 40 is under his arm.

Eastern front, somewhere east of Voronesh, late summer 1942. The motorcycle riflemen of the advance unit of the 24th Panzer Division move out to attack a suspicious, isolated farm along their line of march. The men carry folding shovels in open leather holders on the left side of their belts.

The advance unit has been fired on from the lonely farm, and the drivers of the heavy BMW R 75 1940-44, 26 HP, 750 cc OHV cycles take cover.

The drivers of the heavy BMW machines bring their vehicles back to safety while the motorcycle riflemen carry the attack forward against the farm.

Eastern front, autumn 1942: With his MG 34 machine gun pointed forward on the sidecar and ready to fire. The cycle is a heavy BMW R 75 1940-44, 26 HP, 750 cc OHV type with (driven) sidecar.

Autumn 1942: A transport train rolls toward its station. The vehicles are painted with tan *Afrika-Korps* camouflage paint. In front is a heavy BMW R 12 1935-41, 20 HP, 750 cc SV cycle.

On the way to the eastern front, autumn 1942: Lashing and blocking the BMW R 61 1938, 18 HP, 600 cc SV cycle on a "Stuttgart" type railroad flatcar with stakes for shipment.

Eastern front, autumn 1942: A military police patrol, known as "chained dogs", with a heavy BMW R 12 1935-41, 20 HP, 750 cc SV cycle. Their motorcycle coats have police ring collars. The emblem worn by the policeman in the sidecar and that on the sidecar itself show that they belong to a high command. The "Feldgendarmerie" was a Wehrmacht branch with police duties. A unit in a battalion consisted of three companies, each with three platoons. A company normally had four officers and 22 men with 22 cars, 8 trucks and 28 motorcycles. The police units were part of divisions or larger units. All police ranks were fully trained infantrymen and wore infantry uniforms with special insignia. These two belong to an army corps.

Eastern front, autumn 1942: A messenger in a motorcycle coat on a heavy BMW R 12 1935-41, 20 HP, 750 cc SV cycle with a big leather pouches.

Eastern front, autumn 1942: A formation of 4.7 36 (t) antitank guns on Renault 35 R (f) chassis blocks the way of a heavy BMW R 12 1935-41, 20 HP, 750 cc SV cycle.

Eastern front, southern sector, 1942: Advance unit of a motorcycle rifle company with heavy BMW R 12 1935-41, 20 HP, 750 cc SV cycles prepare for action.

North Africa, autumn 1942: A heavy Zündapp machine on a desert road. The driver of a paratroop unit has brought three soldiers equipped for marching. They wear tan uniform caps of light duck canvas, a khaki *Afrika-Korps* coat and brown variants of the usual army coat, combat packs, musette bags, utensils and flasks. The man on the spare wheel also has a triangular tent panel (Zeltbahn) rolled up as a pack. He also has a steel helmet with tan cloth camouflage cover and canvas boots. The *Afrika-Korps* armband can be seen on the tropical uniform of the officer in the sidecar.

North Africa, autumn 1942: A motorcycle driver adjusts a valve of a heavy BMW R 12 1935-41, 20 HP, 750 cc SV machine.

The driver checks the valve springs on the left cylinder of a heavy BMW R 12 1935-41, 20 HP, 750 cc SV cycle.

Eastern front, autumn 1942: A typical day on the road between Orcha and Smolensk for a heavy BMW R 12 1935-41, 20 HP, 750 cc SV cycle and its crew.

Eastern front, Galicia, winter 1942-43: A heavy BMW R 12 1935-41, 20 HP, 750 cc SV machine. The soldiers wear water-repellent motorcycle coats and "winter head covering." On the front fender is a Soviet gas-mask bag, used as a container.

Voronesh area, early December 1942: A snow-covered heavy BMW R 75, 26 HP, 750 cc OHV cycle with sidecar.

Eastern front, winter 1942-43: A Zündapp KS 750, the competitor to the BMW cycles; on the sidecar is a complete FuG VII radio, with its thin antenna sticking out behind the passenger.

Eastern front, winter 1942-43: A heavy BMW R 75 1940-44, 26 HP, 750 cc OHV cycle with (driven) sidecar. The motorcycle riflemen are wearing winter camouflage suits. In the background is a "Tiger" tank, Type E (88mm KwK 36L/56 gun) with winter camouflage paint.

Eastern front, winter 1942-43: A motorcycle messenger in a winter camouflage suit with a 98k carbine, felt-leather boots and a leather despatch pouch, on a BMW R 35 1937-40, 14 HP, 350 cc OHV medium motorcycle.

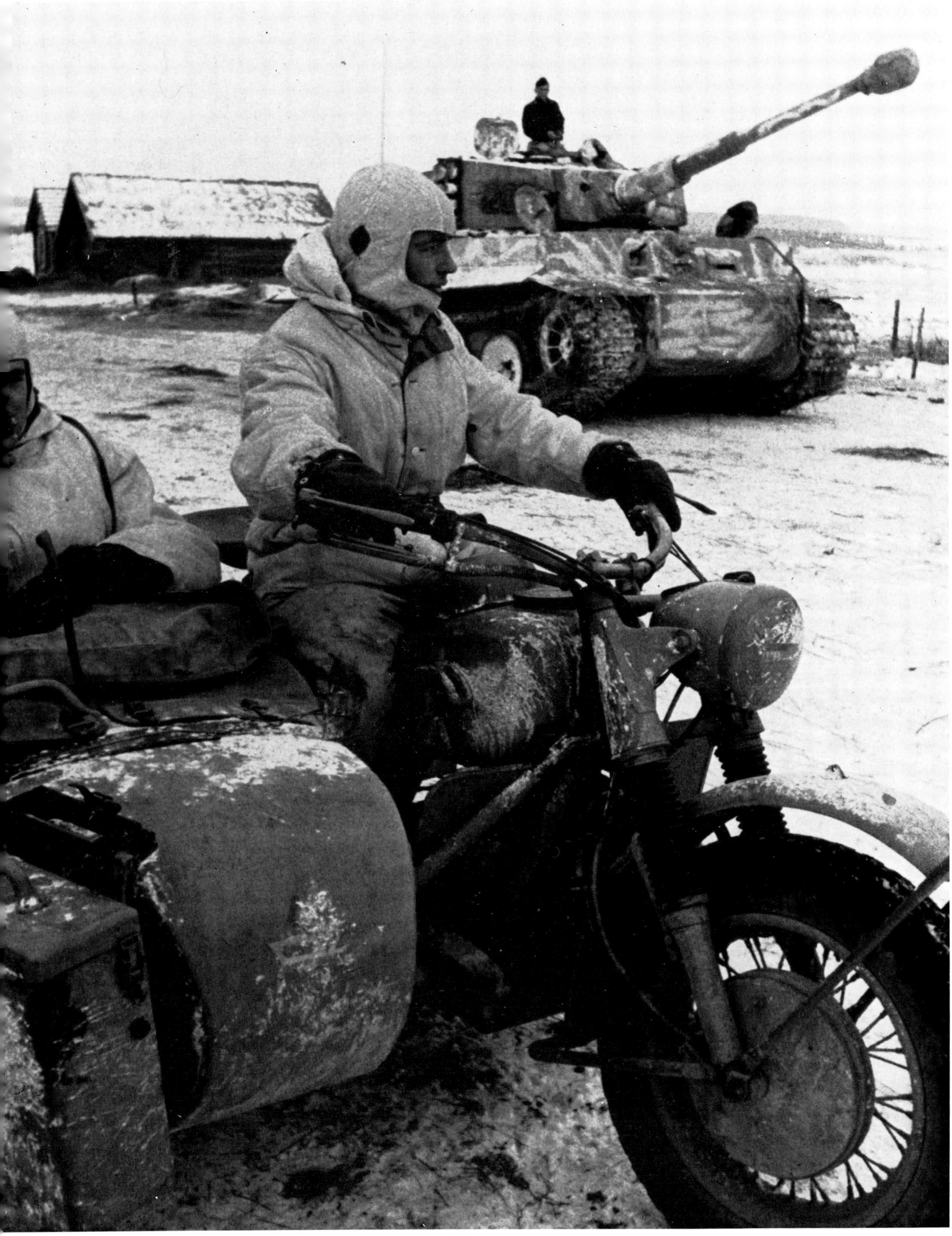

Eastern front, spring 1943: This Zündapp KS 750, stuck in a village street, is being towed out. The two motorcycle riflemen are wearing 1943 model uniform caps and water-repellent motorcycle coats.

Sicily, spring 1943: A heavy BMW R 12 1935-41, 20 HP, 750 cc SV motorcycle.

Balkans, spring 1943: The driver of a BMW R 12 1935-41, 20 HP, 750 cc SV machine asks directions. At left is a young Macedonian herder. A 20-liter fuel can is attached to the sidecar.

Palermo, spring 1943: The men of an *Afrika-Korps* propaganda company on a heavy BMW R 75 1940-44, 26 HP, 750 cc OHV cycle with (driven) sidecar. Both soldiers are wearing 1943 model uniform caps, the driver wears a motorcycle coat, the passenger a 1943 model uniform jacket; his motorcycle coat is lying on the sidecar.

Eastern front, spring 1943: On the road in wet weather is a heavy BMW R 12 1935-41, 20 HP, 750 cc SV machine.

Sicily, spring 1943: At a "traffic checkpoint." In the background is a heavy BMW R 12 1935-41, 20 HP, 750 cc SV cycle.

Southern France, spring 1943: A motorcycle paratroop unit mounts their BMW machines for a training session.

Southern France, spring 1943: A paratroop-motorcycle rifle unit on heavy BMW R 75 1940-44, 26 HP, 750 cc OHV cycles with (driven) sidecars during training.

Southern France, spring 1943: A paratroop-motorcycle rifle unit in marching order (training). The lead cycle is a heavy BMW R 75.

North Africa, Tunis, 1943: A destroyed BMW R 75, silent witness to heavy fighting.

Florence, summer 1943: A motorcycle driver wearing a uniform cap, field gray shirt and shorts poses for a picture on a BMW R 4 1932-36, 12 HP, 400 cc OHV medium sport Reichswehr ("Springer") cycle.

Southern Italy, summer 1943: In front is a heavy BMW R 75 1940-44, 26 HP, 750 cc OHV cycle with (driven) sidecar. On the fuel tank is the helmetlike cover of the engine air filter. At left is an Italian motorcycle rifleman on a Moto Guzzi, in the background a light Italian CV 33-35 tank.

Southern Italy, summer 1943: At left is an Italian CV 33-35 tank with twin machine guns; behind it a heavy BMW R 75 1940-44, 26 HP, 750 cc OHV motorcycle with (driven) sidecar.

Southern Italy, 1943: Motorcycle drivers of a Luftwaffe unit maintaining their heavy BMW R 12 1935-41, 20 HP, 750 cc SV cycle, registration number WL 251269.

WL 251269 gets a cleaning.

The driver cleans the magneto ignition of WL 251269.

The headlight of heavy BMW R 12 cycle WL 251269 gets cleaned.

The last touch and WL 251269 is ready to roll again.

Germany, summer 1943: In marching order and ready to fire, a company of a motorcycle rifle battalion on maneuvers, equipped for the field with 1935 model steel helmets, carrier boxes with gas masks, musette bags, 98 k carbines, and machine guns on the (driven) sidecars of the heavy BMW R 75 1940-44, 26 HP, 750 cc OHV cycles.

The machine gunner of a heavy BMW team. The MG 34 is attached to its holder by a picot arm. The corporal has put the carrying strap of his musette bag on his steel helmet to hold camouflage material.

Mediterranean area, autumn 1943: A heavy BMW R 75 1940-44, 26 HP, 750 cc OHV cycle with (driven) sidecar, belonging to a tank unit. In the foreground is a Panzer IV, Type H tank with a long 75mm gun.

Eastern front, autumn 1943: A motorcycle rifle unit works its way forward with difficulty in rough country.

Eastern front, southern sector, autumn 1943: A motorcycle rifle group advances to check the situation in enemy country. On the back seat of the medium BMW R 35 1932-36, 12 HP, 400 cc OHV cycle is a *Leutnant* (company leader) with an MP 38/40 machine pistol.

Southern Italy, autumn 1943: A motorcycle messenger with a tan steel helmet, 1935 model, and khaki tropic shirt, on a heavy BMW cycle.

Balkans, autumn 1943: An *Obergefreite* and a rifleman of an equipment control staff with their heavy BMW R 12, 750 cc SV cycle with sidecar.

Italy, spring 1944: A heavy BMW R 12 1935-41, 20 HP, 750 cc SV cycle.

Italy, near Rome, June 5, 1944: a motorcycle rifleman of the German rear guard lost his life when his BMW R 75 burned.

France, invasion front, June 1944: Two "chained dogs", as the soldiers not so lovingly called the military police, take a nap on a heavy BMW R 75 1940-44, 26 HP, 750 cc OHV cycle with (driven) sidecar. The two belong to a paratroop division. The driver is wearing an olive green paratrooper combat jacket with the ring collar of the military police, plus a leather despatch pouch and black wartime marching boots. At right on the tank is the hand shift lever, under it the off-road shift lever, near it the upper fork covering, with a carrier box on the left side of the sidecar.

Caen area, mid-June 1944: Three soldiers of a motor-cycle paratroop unit. The driver is wearing a motor-cycle coat and very unusual headgear. The cycle is a heavy BMW R 12, 750 cc SV machine with sidecar.

Normandy, late June 1944: A paratrooper on his heavy BMW R 75 1940-44, 26 HP, 750 cc OHV cycle with (driven) sidecar.

France, Normandy near the English Channel, July 1944: The heavy BMW R 12 1935-41, 20 HP, 750 cc SV machine of a paratroop unit, camouflaged on account of low-flying planes.

France, Normandy, invasion front, July 1944: The *Oberleutnant* of a paratroop unit reads the map at a crossroads. The two soldiers are wearing light camouflage jackets. On the front fork of the heavy BMW R 75 1940-44, 26 HP, 750 cc OHV cycle with (driven) sidecar are the rubber sleeves of the front fork, with the front lifting loop over the fender. Carrying boxes are hung on both sides of the front of the sidecar.

France, Normandy, invasion front, July 1944: A motorcycle advance unit of the paratroopers on duty. The driver is wearing a paratroop steel helmet with a wire net to hold camouflage material, plus a visor.

Italy, summer 1944: An improvised repair shop in the middle of an orange grove. In front is a heavy BMW R 12 1935-41, 20 HP, 750 cc SV motorcycle.

Italy, autumn 1944: The motorcycle riflemen of a paratroop unit in action with their BMW R 75.

France, September 1944: A heavy Zündapp KS 750 machine fleeing from the advancing allied motorized units. The front fender has been removed; the holder for the carriers can be seen on the front of the sidecar.

Hungary, autumn 1944: A lone motorcycle driver on the road near Sopron with a heavy BMW R 12 1935-41, 20 HP, 750 cc SV machine.

Ardennes, Foy Notre Dame, December 29, 1944: No more gasoline. In the midst of an abandoned battery is a medium BMW R 12 1932-36, 12 HP, 400 cc OHV cycle.

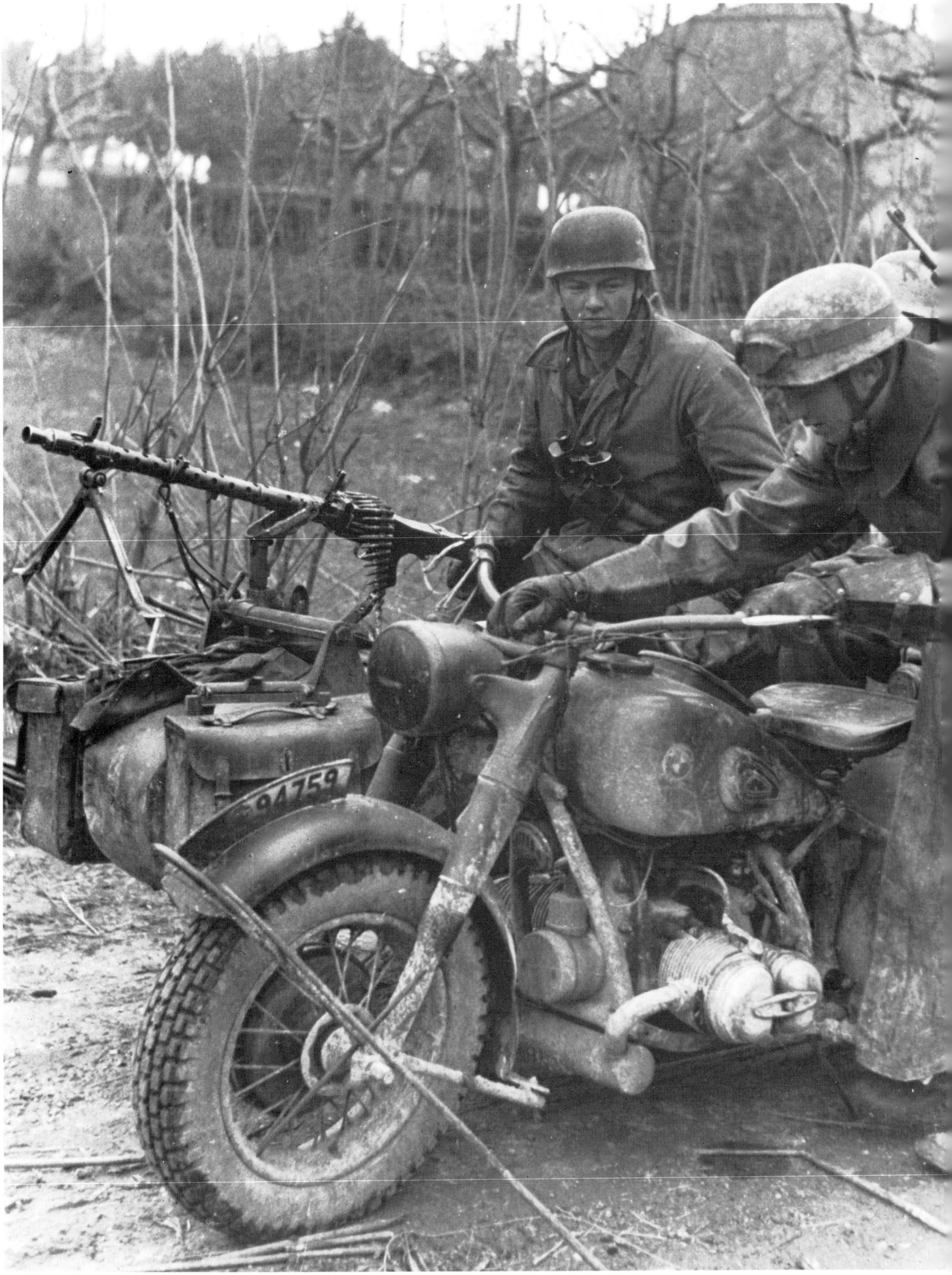
94759

6

TECHNICAL DATA

ON THE MEDIUM AND HEAVY BMW 2-CYLINDER PRODUCTION MOTORCYCLES

1. *R 32 1923-26;*
8.5 HP, 122 kg
A touring machine; 500 cc motor with side valves ("SV"), cylinder longitudinally ribbed, one-piece including cylinder head; BMW two-slide carburetor (for air and gas), three-speed transmission and rear-wheel drive with grease filling, wedge-block rear brake and internal drum front ected by couplings, speedometer drive from the front wheel to the speedometer on the fork bridge. The fuel tank is under the upper frame section. Driveshaft and drive pinion in one piece. Stub axle for the rear wheel, 26" x 3" knobby tires. All subsequent models to the R 11 series II have a BMW two-slide carburetor in various versions. Wedge-block rear brake.

2. *R 37 1925-26;*
16 HP, 134 kg
Chassis as R 32 but now with sport motor with dropped valves in the head ("OHV") and rods. Cylinders crossribbed with separate cylinder heads and valve covers. Wedge-block rear brake.

3. *R 42 1926-28;*
12 HP, 126 kg
A touring machine with 5oo cc ("OHV") SV motor. This motor has cylinder heads, BMW two-slide carburetor, new three-speed transmission with grease filling, speedometer drive from the rear gearbox flange to the speedometer on the fuel tank. Additional innovations as compared to the two earlier machines include a driveshaft brake with thin shoes mounted on the gearbox, plus rear brake filled with motor oil; sidecar attachments provided. 26 x 3.5" ND or 27 x 2.75" HD tires. Identifying marks as opposed to R 32 and R 37 are the thin driveshaft brake and the speedometer on the fuel tank.

4. *R 47 1927-28;*
18 HP, 130 kg
Same chassis as R 42, but 500 cc OHV sport motor.

Italy, spring 1945: Motorcycle riflemen of a paratroop unit with the machine gun on the (driven) sidecar of their heavy BMW R 75 1940-44, 26 HP, 750 cc OHV cycle ready to fire.

5. R 52 1928-29;
12 HP, 152 kg
A 500 cc SV touring machine, long-stroke motor, gearbox with new oil filler and speedometer drive from the gearbox flange to the speedometer on the fuel tank. Larger front-wheel brake. 26 x 3.5" ND or 26 x 3.25" HD tires, for the sidecar 27 x 4" ND. The frame is the same in principle as the R 42-R 47 type. Essential differences from previous models are the enlarged front brake and the shift-lever linkage on the motor housing.

6. R 57 1928-30;
18 HP, 150 kg
Same as R 52 but with a new 500 cc OHV sport motor.

7. R 62 1928-1929;
18 HP, 155 kg
First 750 cc SV motor; chassis as R 52-R 57.

8. R 63 1928-29;
24 HP, 155 kg
First 750 cc OHV motor. Chassis as R 52-R 57. Essential difference is only the new motor. All types listed to this point were produced without headlights and directional signals; these were offered at extra cost as special equipment.

9. R 11 Series I, 1929-30;
18 HP, 162 kg
750 cc SV motor. The pressed-steel frame was introduced with this model, providing greater rigidity — because of the higher motor output. Essential parts of the front fork were also made of pressed steel. Front and rear stub axles for the first time. Electric lighting system with magneto ignition and horn now standard. Thin driveshaft brake, shaft and pinion still in one piece. Engine-gearbox block of the R 62. Fuel tank on the upper frame member.

R 11 Series II, 1930-31;
18 HP, 162 kg
Improved and widened driveshaft brake, minor improvements to the chassis.

R 11 Series III, 1931-33;
18 HP, 162 kg
Differences from Series I and II: Motor with heater duct to intake tube, for the first time a Sum carburetor with damp-air filter. Theft proofing, fuel strainer in fuel pump. Divided driveshaft from this model on; the drive pinion has its own bearings. Considerably decreased running noise and durability.

R 11 Series IV, 1933;
18 HP, 162 kg
With this type the wing shifting on the frame was introduced. Saddle with pull springs (previously pressure springs), friction shock absorbers on the front fork and other minor improvements. Same motor as in Series III.

R 11 Series V, 1934;
20 HP, 162 g
New motor with two Amal carburetors, each with one damp air filter. Battery ignition for the first time, plus starting via one chain each for the driveshaft and battery ignition (all previous models had gear drive). The tires of Series I-V were the same as those of the R 52 and R 63. With sidecar drive, optional 27 x 4" tires at extra charge.

10. R 16 Series I and II, 1929-32;
25 HP, 165 kg
Same chassis as R 11 Series I and II, but 750 cc OHV sport motor like that of R 63.

R 16 Series III, 1932;
33 HP, 165 kg
With sport motor, two carburetors and damp air filter, otherwise as R 11 Series III.

R 16 Series IV, 1933;
33 HP, 165 kg
Same motor as R 16 Series III, but different compression. Same chassis as R 11 Series IV.

R 16 Series V, 1934;
33 HP, 165 kg
Same chassis as R 11 Series V, but OHV sport motor, basic design as R 11 Series V.

11. *R 12 1935-38;*
18 or 20 HP, 162 kg
Built in one- and two-carburetor versions; new chassis, first use of telescopic fork with hydraulic shock absorbers. Compared to the previous short-arm pinion with leaf springs and its many joints, the new fork is practically maintenance-free, has a longer life, long-stroke springs and thus considerably better roadholding. Further improvements: Four-speed transmission and shift linkage on the frame, first use of interchangeable wheels and speedometer mounted on the headlight, steel-belted 3.5 x 19" tires. The motor (750 cc SV) is like that of the R 11 Series V, as is most of the frame, but the rear fender folds up; the front fender goes farther down, providing increased dirt protection. From this model on, there is no longer a fuel filter in the tank. The one-carburetor model was built for the Wehrmacht longer than indicated above.

12. *R 17 1935-37*
33 HP, 165 kg
750 cc OHV sport motor like that of R 16 Series V; chassis and gearbox like R 12; better spray protection.

13. *R 5 1936-37;*
24 HP, 165 kg
Ground-breaking new design: 500 cc OHV motor with tunnel housing, two camshafts with chain drive, two Amal carburetors, hairpin valve springs, generator and battery ignition. Four-speed gearbox with two shafts and anchor foot shift. Slip hooks and small auxiliary hand shift lever. Return to tube frame but now welded; closed triangular double frame of conical steel tubes. Saddle tank, rubber swinging saddle, adjustable as to softness and height. From this model on, all two-cylinder types had footrests instead of running boards and so-called heel brakes. Improved telescopic fork with greater spring length and adjustable hydraulic damping. Interchangeable wheels, front and rear stub axles. For the first time, front-wheel kick-stand, plus fender brace.

14. *R 6, 1937;*
18 HP, 175 kg
Chassis like R 5, long-stroke 600 cc SV motor, but with one carburetor, one driveshaft and bevel gear drive. Central damp air filter on the gearbox.

15. *R 51, 1938 — certainly built to end of 1939;*
24 HP, 182 kg
Corresponds in principle to the R 5 and R 6, but with rear-wheel suspension, new rear-axle drive and cross link. Front fork like R 5, but damping no longer adjustable. 500 cc OHV motor with two camshafts and chain drive like R 5. Central damp air filter and gearbox like R 6.

16. *R 61 1938 — certainly built to end of 1939;*
18 HP, 184 kg
600 cc SV motor.

17. *R 66 1938 — certainly built to end of 1939;*
30 HP, 187 kg
600 cc OHV motor.

18. *R 71 1938 — certainly built to end of 1939;*
22 HP, 187 kg
750 cc SV motor. All three previous types have the same chassis and gearbox as R 51. But the motors have bevel-gear drive for the driveshaft and generator.

19. *R 75, 1941;*

26 HP, 420 kg (including sidecar)
750 cc OHV motor. Specially developed as off-road vehicle with driven sidecar wheel. Special design features: Gearbox with 4 road and 4 off-road gears, 2 reverse gears; locking differential for sidecar drive, hydraulic brakes, first use of automatic ignition timing adjustment. From this model on, all subsequent models had automatic regulation of ignition timing. The generator is attached to the front driveshaft mount.

The front wheel of a heavy BMW R 12 1935-41, 20 HP, 750 cc SV cycle with winged fender, wheel fork, brake drum, spokes, rim and 350-19 off-road tires, Continental brand.

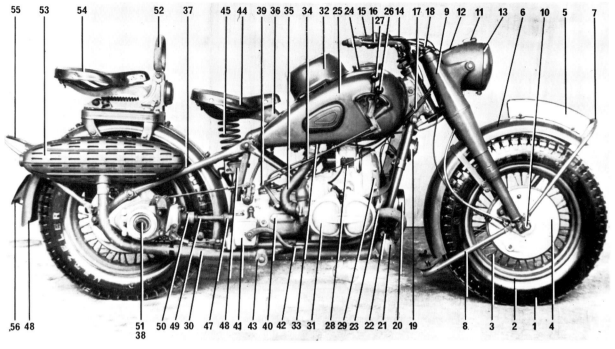

Motorcycle, right side:
1. Off-road tires, Metzeler 120-16, 1.7 atm.
2. Wheel rim
3. Spokes
4. Front brake drum
5. License plate
6. Front fender
7. Front lifting loop
8. Front kickstand
9. Front wheel fork
10. Stub axle
11. Light switch
12. Speedometer with odometer
13. Headlight
14. Handlebars
15. Accelerator handle
16. Hand brake lever
17. Brake and clutch cables
18. Steering bearing
19. Horn
20. Generator
21. front muffler
22. Front sidecar attachment
23. Sidecar heater attachment
24. Fuel filler cap
25. Fuel tank
26. Off-road shift lever
27. Hand shift lever
28. Magneto
29. Chain cover
30. Exhaust pipe
31. Cylinder head cover
32. Air filter in fuel tank
33. Rubber coupling with hose clamps
34. Shift linkage
35. Fuel intake line
36. Saddle brace (saddle carrier)
37. Rear frame
38. Rear sidecar attachment
39. Shift lever for differential lock
40. Differential
41. Brake fluid reservoir
42. Foot brake pedal
43. Footrest
44. Driver's saddle
45. Saddle spring
46. Safety box.
47. V-belt drive
48. Rear fender
49. Driveshaft
50. Axle stay
51. Sidecar drive and connection
52. Handhold
53. Rear muffler
54. Passenger seat (or luggage rack)
55. Rear lifting loop
56. Rear license plate

Sidecar of a BMW R 75, right side:
1. Spare wheel
2. Luggage space
3. Sidecar fender
4. Taillight
5. Sidecar light
6. Tire (Continental)
7. Wheel rim
8. Spokes
9. Brake drum
10. Sidecar body
11. Handhold
12. Pack holder

The luggage space of the BMW R 75 army sidecar with spare wheel.

Heavy BMW R 75 1940-44, 26 HP, 750 cc OHV cycle with (driven) sidecar, of a paratroop unit. The MG 34, standard German infantry weapon, is mounted on a standard with swinging arm. On the sidecar is a pack wrapped in a triangular tent canvas.

Side view of the BMW R 75 1940-44, 26 HP, 750 cc OHV cycle with (driven) sidecar of a paratroop unit. On the sidecar is an MG 34 (7.92 mm) with 34 belt drum.

s. KRAFTRAD BMW R 75 750 ccm

MOTORCYCLE BMW R 75 750 cc
Telescopic front wheel fork

Teleskop-Vorderradgabel

Die BMW-Vorderradgabel erfüllt alle an eine einwandfreie Federung des Vorderrades gestellten Ansprüche, wie geringe ungefederte Massen, gute Bodenhaftung des Vorderrades und stoßfreies Fahren, sowie staub- und schmutzsichere Unterbringung aller beweglichen Teile in einer formschönen Verkleidung.

Die Vorderradgabel besteht aus zwei kräftigen Holmen, welche durch ober- und unterhalb des Lenkkopfes angebrachte Querverbindungen (die Lenkerplatte 1 und die Gabelführung 3) verschraubt sind. Jede Gabelhälfte besteht aus einem feststehenden Führungsrohr (dem Gabelrohr 2), über welches unten das Gleitrohr (Gabelendstück 13) aufgeschoben ist. Die Führung geschieht durch die obere 10 und untere 12 Führungsbuchse, die Abdichtung durch den druckfesten Dichtring 9. Die Abstützung des Gabelrohres erfolgt durch eine Schraubenfeder 4, deren unteres Ende im Einspannstück 8 des Gabelendstückes 13 und deren oberes Ende im Einspannstück der Gabelführung 3 eingeschraubt ist, so daß die Feder sowohl auf Druck, als auch auf Zug beansprucht werden kann, also die Fahrstöße und die Rückschläge aufnehmen kann.

Ein doppeltwirkender Ölstoßdämpfer in jeder Gabelhälfte dämpft die Fahrschwingungen. Der Stoßdämpfer besteht aus einer in die Verschlußschraube der Gabel geschraubten Stoßdämpferstange 5, welche am unteren Ende das Stoßdämpferventil 11 trägt, aus dem am Gabelendstück 13 befestigten Stoßdämpferrohr 6 und der Führungsbuchse 7, welche zusammen mit der Stoßdämpferstange 8 die ringförmige Stoßdämpferdüse 7 bildet. Unten im Stoßdämpferrohr 6 befindet sich das Bodenventil 14, so daß aus dem Raum außerhalb des Stoßdämpferrohres Öl nachgesaugt werden kann.

Die Wirkungsweise des Stoßdämpfers ist folgende:
Beim Durchfedern der Gabel bewegt sich das Gleitrohr (Gabelendstück 13) und damit das Stoßdämpferrohr 6 nach oben, das untere Ende der Stoßdämpferstange 5 wirkt dabei wie ein Verdrängerkolben (Plungerkolben) und verdrängt einen Teil des im Stoßdämpferrohr 6 befindlichen Öles. Dieses Öl muß, weil das Bodenventil 14 geschlossen bleibt, am Stoßdämpferventil 11 vorbei nach oben. Das verdrängte Öl hat über dem Stoßdämpferventil 11 nicht Raum genug, das überschüssige Öl wird deshalb durch die Stoßdämpferdüse 7 gepreßt, wodurch die Dämpfung erzielt wird. Das durchgepreßte Öl fließt oben über den Rand des Stoßdämpferrohres 6 und sammelt sich unten im Gabelendstück 13. Beim Rückschlag (bewirkt durch die sich entspannende Schraubenfeder 4) bewegt sich das Gabelendstück 13 und damit das Stoßdämpferrohr 6 wieder abwärts. Dabei schließt sich das Stoßdämpferventil 11 und es wird Öl durch das Bodenventil 14 angesaugt. Das oberhalb des Stoßdämpferventiles 11 befindliche Öl wird wiederum durch die Stoßdämpferdüse 7 nach oben gepreßt und dämpft damit auch den Rückschlag (doppelte Wirkung!).

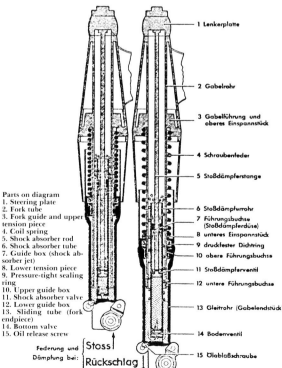

Parts on diagram
1. Steering plate
2. Fork tube
3. Fork guide and upper tension piece
4. Coil spring
5. Shock absorber rod
6. Shock absorber tube
7. Guide box (shock absorber jet)
8. Lower tension piece
9. Pressure-tight sealing ring
10. Upper guide box
11. Shock absorber valve
12. Lower guide box
13. Sliding tube (fork endpiece)
14. Bottom valve
15. Oil release screw

Federung und Dämpfung bei: Stoss / Rückschlag

Lower left:
Suspension and damping at: pressure, recoil

1 Lenkerplatte
2 Gabelrohr
3 Gabelführung und oberes Einspannstück
4 Schraubenfeder
5 Stoßdämpferstange
6 Stoßdämpferrohr
7 Führungsbuchse (Stoßdämpferdüse)
8 unteres Einspannstück
9 druckfester Dichtring
10 obere Führungsbuchse
11 Stoßdämpferventil
12 untere Führungsbuchse
13 Gleitrohr (Gabelendstück)
14 Bodenventil
15 Ölablaßschraube

The BMW front wheel fork meets all the demands placed on a trouble-free front wheel suspension, such as small unsprung weight, good front-wheel roadholding and jolt-free riding, plus dust- and dirt-free mounting of all moving parts in a good-looking cover.

The front-wheel fork consists of two strong uprights which are united by crosspieces above and below the steering head (steering plate 1 and fork guide 3). Each half of the fork consists of a fixed guide tube (fork tube 2), over which the sliding tube (fork endpiece 2) is placed from below. Guiding is done by the upper 10 and lower 12 guiding boxes, sealing by the pressure-fast sealing ring 9. The support of the fork tube is done by a coil spring 4, its lower end screwed into tension piece 8 of the fork endpiece 13 and its upper end into the tension piece of the fork guide, so that the spring can be activated by pressure as well as pulling and thus can take up the road bumps and recoils.

A double-acting oil-pressure shock absorber in each half of the fork absorbs driving swinging. The shock absorber consists of a shock absorber rod 5 screwed into the closing screw of the fork and carrying the shock absorber valve 11 on its lower end, out of the shock absorber rod 6 and guiding box 7 fastened to the fork endpiece 13; along with the shock absorber rod 8, they form the ring-shaped shock absorber jet 7. At the bottom of the shock absorber rod 6 is the bottom valve 14, which allows oil to be sucked out of the space outside the shock absorber rod.

The operation of the shock absorber is as follows: In compression of the fork, the sliding tube (fork endpiece 13) and thus the shock absorber rod 6 move upward, the lower end of the shock absorber rod 5 thus works like a compressor piston (plunger piston) and compresses part of the oil in the shock absorber tube 6. This oil must, since the bottom valve 14 stays closed, go past shock absorber valve 11 and upward. The compressed oil does not have enough room over the shock absorber valve 11, therefore the excess oil is pressed through the shock absorber jet 7, which produces the damping. The oil pushed through flows up through the rim of the shock absorber tube 6 and collects down in the fork endpiece 13. In the recoil (caused by the tensing coil spring 4) the fork endpiece 13 and therefore also the shock absorber tube 6 move down again. Thus the shock absorber valve 11 is closed and oil is sucked through the bottom valve 14. The oil above the shock absorber valve 11 is then pushed upward through the shock absorber jet 7 and thus also dampens the recoil (double action!).

LEHRTAFEL 14 u. 18 (ERLÄUTERUNGEN)   BAYERISCHE MOTOREN WERKE AG., MÜNCHEN

The telescopic fork of the BMW R 75. Notice the big mechanical inboard drum brake with a drum diameter of 250 mm.

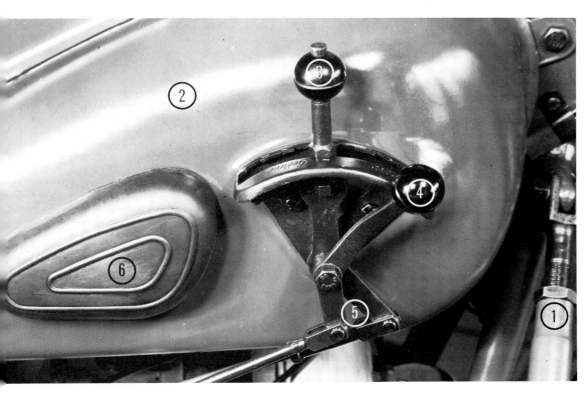

1. Sidecar adjustment
2. Fuel tank
3. Hand shift lever
4. Off-road shift lever
5. Shift linkage
6. Knee pad

1. Wing screw for the steering damper
2. Fuel tank
3. Fuel filler cap
4. Off-road shift lever
5. Hand shift lever
6. Sheet metal cowling of the fuel air filter

Fuel air filter of the BMW R 75: Through the inlet openings under the sheet metal cowling comes inducted air into the accordion-shaped felt bellows and then through a tube to the two intake tubes.

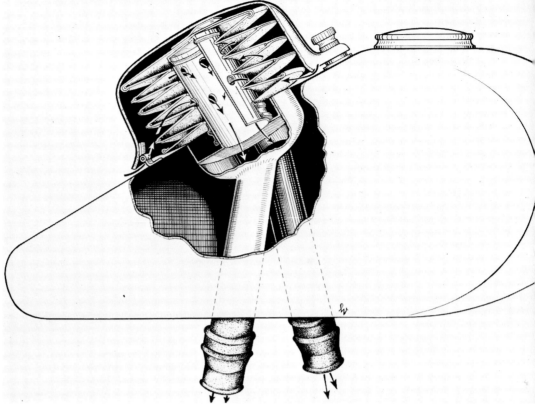

By raising the cowling, the felt bellows filter is accessible and can easily be removed to shake out the dust.

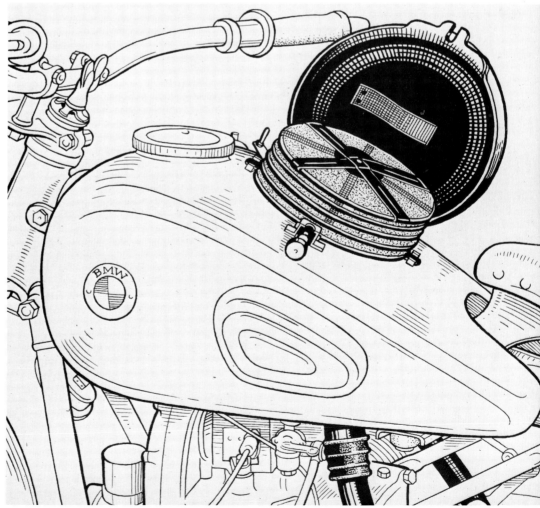

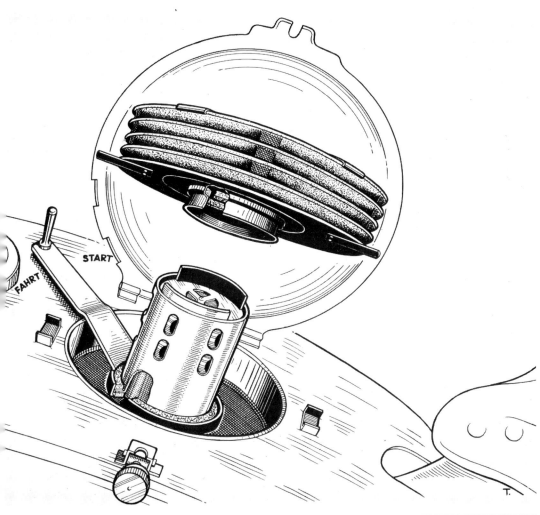

To clean the felt bellows filter it is necessary to fold up the protective cowling, exposing the filter bellows which can be removed. Shaking out the felt filter is sufficient to make the machine ready to drive again immediately. The lever arm serves to adjust the inlet opening, so that when the motor is started, a richer fuel mixture is created. After warming up the motor, this lever is pushed back to "drive" and the air intake louvers and opened completely.

Cleaning the air filter of a BMW R 75: A mechanic shakes out the accordion-shaped felt bellows filter. Before the raised sheet metal cowling are the openings to the air pipe.

SCHWERES KRAFTRAD 750 cm

mit Seitenwagen (angetrieben)

BMW Typ R 75

1. Kundendienstdurchsicht

(nach 500 - 1000 km Fahrstrecke)
Ausführungshinweise siehe D 605/5
(Gerätbeschreibung und Bedienungsanweisung)

1. Ölwechsel im Motor (siehe Rückseite).
2. Ölwechsel im Getriebe (siehe Rückseite).
3. Ölwechsel im Radantrieb (siehe Rückseite).
4. Ölwechsel im Seitenwagenantrieb (siehe Rückseite).
5. Ölwechsel in beiden Gabelhälften (siehe Rückseite).
6. Abschmieren laut Schmierplan.
7. Allgemeine Schrauben- und Mutternkontrolle (Achsen, Schutzbleche, Gepäckträger, Lenker, Auspuff, Saugleitung, Vergasermuttern, Zylinder, Zylinderköpfe Motor und Seitenwagen)
8. Lenkungseinstellung prüfen (muß vollkommen spielfrei sein, Lenker muß aber i aufgebocktem Zustand nach jeder Seite frei fallen)!
9. Ventilspiel prüfen (bei kaltem Motor 0,25 mm, im Kolonial- und Gebirgsdiens 0,30 mm)
10. Vergaser, Kraftstoffhahn und Luftfilter reinigen
    (Füllung im Ölsumpf erneuern — 40 cm³)
11. Drosselstift nach Markierung kürzen. Leerlauf wenn nötig nachstellen.
12. Bremswirkung prüfen (Handbremse, Hinterrad- und Seitenwagenbremse), Brems flüssigkeitsstand prüfen und ergänzen, falls erforderlich Öldruckbremse en lüften.
13. Kupplungshebel einstellen (Spiel am Lenker etwa 10 mm).
14. Sammler prüfen, evtl. destilliertes Wasser nachfüllen.
15. Reifendruck prüfen (siehe Rückseite).
16. Befund fahren.

Arbeitspreis: RM. 12,-

Fahrgestell-Nr.:.................................... Motor-Nr.:................................

Die Arbeit wurde nach .... km ausgeführt.

...., den........................19........

....
(Stempel und Unterschrift)

 BAYERISCHE MOTOREN WERKE Aktiengesellschaft, München

BMW — M 2744 VK 388/516 5,0 10.41. Br./0024

Reifendruck:	Vorderrad	1,75 atü
	Hinterrad	2,75 atü
	Seitenwagenrad	1,75 atü

Schmiermittel:

Motor:	Motoreneinheitsöl der Wehrmacht	2 Ltr.
Ölsumpf im Luftfilter:	Motoreneinheitsöl der Wehrmacht	40 cm³
Getriebe:	Motoreneinheitsöl der Wehrmacht	1,250 Ltr.
Radantrieb:	Gargoyle Mobiloel Epwi oder ein anderes vertraglich zugelassenes Getriebeöl	0,100 Ltr.
Gabel:	Motoreneinheitsöl der Wehrmacht	je Hälfte 0,160 Ltr.
Übrige Schmierstellen (Radnaben usw.) abschmieren lt. Schmierplan:	Vertraglich zugelassenes Abschmierfett	

HEAVY MOTORCYCLE 750 CC with sidecar (driven) BMW Type R 75

1. Customer service procedure (after 500-1000 km)
For procedural guidelines see D 605/5
(Equipment description and service instructions)

1. Change oil in motor (see reverse).
2. Change oil in gearbox (see reverse).
3. Change oil in wheel drive (see reverse).
4. Change oil in sidecar drive (see reverse).
5. Change oil in both halves of the fork (see reverse).
6. Lubricate according to lubrication diagram.
7. Check all screws and bolts (axles, fenders, luggage rack, steering, exhaust, intake, carburetor screws, cylinders, cylinder heads, motor and sidecar).
8. Test steering system (must be completely play-free, handlebars must fall freely to either side when up on blocks(!
9. Test valve play (with cold motor 0.25 mm, in rural and mountain service 0.30 mm).
10. Clean carburetor, fuel pump and air filter (change filling in oil sump — 40 cc).
11. Shorten choke bar as marked. Adjust free action if necessary.
12. Test brake effect (hand brake, rear wheel and sidecar brake), check brake fluid level and refill, if necessary, let air out of hydraulic brake.
13. Adjust clutch lever (play of lever about 10 mm).
14. Test battery, add distilled water if necessary.
15. Check tire pressure (see reverse)
16. Test-drive.

Pages 170 and 171: The original of the first customer service instructions of October 1941.

Technical Data of the BMW R 75

Motor:

Operation	Four-stroke
Stroke	78 mm
Bore	78 mm
Number of cylinders	2
Displacement	745 cc
Compression ratio	1 : 5.6-5.8
Sustained power	26 HP at 4400 rpm
Torque	5 mkg at 3600 rpm
Piston play	0.07-0.08 mm
Valve play	0.25 mm for intake and exhaust valves with cold motor (0.30 mm for rural and mountain service)
Valve times	(adjust to Inlet opens 16 degrees before upper 0.6 mm valve play) dead point, closes 24 degrees after lower dead point. Exhaust opens 24 degrees before lower dead point, closes 16 degrees after upper dead point.
Ignition system	Noris ZG 2a magneto
Ignition setting time setting	Automatic centrifugal ignition
Ignition adjustment	At upper dead point (self-acting adjustment to 35 degrees)
Spark plugs	Bosch W 175 T 1
Electrode gap	0.5-0.6 mm
Generator	6 volt, 50-70 watt (voltage regulator) (Noris, Type DS 6/50)
Battery	7 amperes per hour (Ah)
Fuses	2 15-ampere
Cooling	Air cooling
Motor lubrication	Circulation
Oil pump	Geared pump
Oil consumption	Normally 0.4 to 1 liter per 1000km
Carburetor	Right Graetzin Sa 24/1, left Greatzin 24/2, needle jet 42, main jet 100, idle jet 35, needle setting I, air regulator screw for idle 1.5-1.75 turns outward

Chassis:

Frame	Multiple bolted tube frames
Fork	2 telescoping fork halves
Springs	2 coil springs (in fork)
Shock absorbers	Double-acting hydraulic (in fork)
Clutch	One-plate dry clutch
Gears	BMW gears with built-in off-road transmission

Lever set to:

	road	off-road
Forward speeds	4	4
Reverse speeds	1	1
Ratios: 1st gear	3.22	4.46
2nd gear	1.83	2.54
3rd gear	1.21	1.67
4th gear	0.90	1.24
Reverse	2.41	3.30
Starter crank	2.91	4.03

Maximum speeds:

1st gear	22 kph	14 kph
2nd gear	44 kph	24 kph
3rd gear	66 kph	42 kph
4th gear	95 kph	65 kph

Transmission from gearbox	Driveshaft with rubber crosslink to driven wheels
Wheel drive ratio	6.05 (to chassis #764056: 5.69)
Differential	bevel gear (lockable)
Wheels driven	Rear and sidecar wheels

Suspension:

Front wheel	Coil springs (in fork)
Sidecar wheel	Tube (works like torsion bar)
Sidecar body	Leaf springs
Saddle	Swinging saddle

Brakes:

Foot brake	Hydraulic, on rear and sidecar wheels
Hand brake	Cable, on front wheel
Wheels	Spoked
Rim size	3.00 D x 16
Tire size	4.50 - 16 (off-road tread)
Air pressure	Front and sidecar 1.75, rear 2.75 atm.
Steering	Handlebars
Steering head bearing	Shoulder bearing (2 x 20 balls with 6.5 mm diameter)
Toe-in (unladen)	2-3 mm
Lead (sidecar)	0-10 mm
Trail (front wheel)	45 mm
Wheelbase	1444 mm
Track	1180 mm

Vehicle:

Length (with sidecar)	2400 mm
Width (with sidecar)	1730 mm
Height	1000 mm
Turning circle to left	4.7 meter diameter
to right	3.6 meter diameter
Bound clearance	275 mm
Undertray clearance	150 mm
Weight ready to drive	420 kg (with sidecar)
Allowable gross weight	840 kg (with sidecar)
Lowest sustained speed	3 kph
Autobahn speed	80 kph (fully laden)
Top speed	95 kph
Climbing, fully laden:	
Short climb	40 degrees
Long climb	35 degrees
Fording ability	350 mm
Fuel consumption (road)	6.3 liters per 100 km at 63 kph
Range	380 km
Fuel consumption (off-road)	up to 8.5 liters per 100 km
Capacities:	
Fuel tank	24 liters (3 liters reserve)
Motor oil: Motor	2 liters
Gearbox	.25 liters
Fork	0.16 liters per half of fork
Oil sump in air filter (only for damp air filter)	0.04 liters
Oil: Wheel drives	0.30 liters
Sidecar drive	0.10 liters

Engine block of the BMW R 75 1940-44, 26 HP, 750 cc OHV.

s. KRAFTRAD BMW 750/275 (R 75)

Motor

Motor

The crankcase (16) is cast in one piece of aluminum. The front part of the crankcase is formed as a valve gear cover and closed by an aluminum cover, the casing cover (23). In the back part is the swinging disc (14) with the clutch. The cylinders (42) are made of cast iron, cylinder heads (35) and covers (29) of aluminum. The crankshaft (9) is set at 180 degrees on its coupling and runs in three bearings. Both connecting rods (63) are attached to the crankshaft by roller bearings. The piston bolts secured by spring rings move in bronze liners. The aluminum pistons each bear two compression (53) and oil-scraping rings (54 and 56). The clutch (12) is a single-disc dry clutch.

The motor has overhead dropped valves (34). These are activated by the cams of the camshaft (19) via tappets (49), pushrods (37) and rockers (32). The camshaft (19) runs in ball bearings activates the rotary changer of the de-aerator (22) connected to the drive wheel. All the control gears are diagonally toothed.

The gear-driven oil pump (45) is driven by gears from the crankshaft (44). The pump sucks the lubricating oil out of the oilpan (43) and pushes it through boreholes in the oil-spreading rings (41) that supply the crankshaft bearings. The dispersed oil reaches the pistons (39), piston bolts and camshaft (19). The control gears are additionally supplied with oil through a special duct (47). The rockers (32) and valves (34) in the cylinder heads (35) receive the lubricating oil through the protective tubes (36). The excess oil flows through the oil return pipes (64) and back into the oilpan (43).

The ignition power is produced in the magneto (17). The voltage-regulating generator (50) is the source of power for the lights and horn. The motor and gearbox are bolted together to firm a single powerplant block.

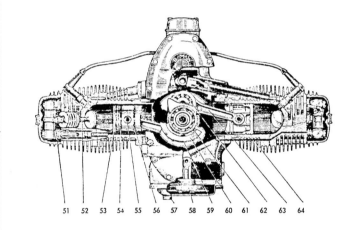

Das Kurbelgehäuse (16) ist aus Leichtmetall in einem Stück gegossen. Der vordere Teil des Gehäuses ist als Steuerräderkasten ausgebildet und mit einem Leichtmetalldeckel, dem Steuerkastendeckel (23) abgeschlossen. Im hinteren Teil befindet sich die Schwungscheibe (14) mit der Kupplung. Die Zylinder (42) sind aus Grauguß, Zylinderköpfe (35) und -deckel (29) aus Leichtmetall. Die Kurbelwelle (9) ist an den Kurbelzapfen um 180° versetzt und läuft in drei Kugellagern. Beide Pleuelstangen (63) sind auf der Kurbelwelle rollengelagert. Die durch Sprengringe gesicherten Kolbenbolzen gleiten in Pleuelbuchsen aus Bronze. Die Leichtmetallkolben tragen je zwei Verdichtungs- (53) und Ölabstreifringe (54 und 56).
Die Kupplung ist eine Einscheibentrockenkupplung (12).
Der Motor ist obengesteuert und hat hängende Ventile (34). Diese werden durch die Steuernocken der Nockenwelle (19) über Stößel (49), Stoßstangen (37) und Kipphebel (32) gesteuert. Die Nockenwelle (19) läuft in Kugellagern und betätigt den mit dem Antriebsrad gekuppelten Drehschieber des Entlüfters (22). Sämtliche Steuerräder sind schrägverzahnt.
Die Zahnrad-Ölpumpe (45) wird über Stirnräder von der Kurbelwelle (44) angetrieben. Die Pumpe saugt das Schmieröl aus der Ölwanne (43) und drückt dasselbe durch Bohrungen in die Ölschleuderringe (41), welche die Lagerstellen der Kurbelwelle versorgen. Das abgeschleuderte Öl gelangt an den Kolben (39), Kolbenbolzen und die Nockenwelle (19). Die Steuerräder werden durch eine gesonderte Leitung (47) zusätzlich mit Öl versorgt. Die Kipphebel (32) und Ventile (34) in den Zylinderköpfen (35) erhalten das Schmieröl durch die Stoßstangenschutzrohre (36). Das überschüssige Öl fließt durch die Ölrücklaufrohre (64) wieder in die Ölwanne (43).
Der Zündstrom wird im Magnetzünder (17) erzeugt.
Die spannungregelnde Lichtmaschine (50) ist die Stromquelle der Lichtanlage und des Horns.
Motor und Getriebe sind zu einem Triebwerkblock verschraubt.

1. Off-road shift lever
2. Distributor gear shift
3. Gearshift lever
4. Clutch lever
5. Air intake duct
6. Sparkplug cable
7. Hose couplings
8. Gearbox
9. Gearbox vent
10. Closing ring
11. Clutch disc
12. Pressure plate
13. Clutch spring
14. Flywheel
15. Distributor and regulator
16. Crankshaft
17. Magneto
18. Holding hook
19. Camshaft
20. Magneto drive gear
21. Camshaft drive gear
22. De-aerator
23. Valve gear cover
24. Voltage regulator
25. Sparkplug cable
26. Brake fluid hose connector
27. Brake fluid reservoir vent
28. Closing screw
29. Cylinder head cover
30. Holding pad
31. Carburetor
32. Rocker
33. Valve spring
34. Valve
35. Cylinder head
36. Protective tubing
37. Bumper
38. Oil release screw
39. Piston
40. Oil strainer
41. Oil dispersal ring
42. Cylinder
43. Oilpan
44. Crankshaft
45. Oil pump
46. Oil pump drive gear
47. Oil jet
48. Drive gear from the crankshaft
49. Tappet
50. Generator
51. Rocker bridge
52. Valve control box
53. Compression ring
54. Oil scraping ring
55. Piston bolt and spring ring
56. Oil scraping ring
57. Oil release screw
58. Oil strainer
59. Crankshaft
60. Oil dispersing ring
61. Crankshaft bearing
62. Connecting rod bearing
63. Connecting rod
64. Oil return from cylinder head

LEHRTAFEL 19 (ERLÄUTERUNGEN) BAYERISCHE MOTOREN WERKE AG., MÜNCHE

1. Rear lifting loop
2. License plate
3. Rear fender
4. Rear muffler with heat shield
5. Exhaust pipe
6. Rear brake drum
7. Spokes
8. Wheel rim
9. 4.50-16 Metzeler tires
10. Wheel drive with differential
11. Saddle spring
12. Frame tube
13. Variable gears
14. Rear kickstand
15. Brake fluid reservoir
16. Footrest

S. KRAFTRAD BMW 750/275 (R 75)

Radantrieb mit Ausgleichgetriebe

Wheel drive with differential gears

To equalize the various turning speeds of the rear and sidecar wheels on curves, a wheel drive with differential gears is built in. The differential is a spur-gear drive with unequal torque division (see also D. "The Torque"). It locks itself in order to prevent one wheel from spinning on muddy roads or off the road.

The drive bevel gear (19, 35 & 38) drives the crown gear (18, 34 & 39) linked with the inner differential housing (10 & 43). The drive torque is divided in the differential, via identical gears, to the rear wheel (drive wheel, 17, 36 & 48) and carrier wheel (14, 33 and 47), to the sidecar wheel (drive wheel, 55) and carrier wheel (46); thus being divided between the rear wheel and sidecar wheel. The spur gear (32 & 50) to drive the rear wheel meshes with the smaller differential spur gears (30 & 40, 31 & 49), while the spur gear (28 & 53) to drive the sidecar wheel via the intermediate differential gears (27 & 42, 29 & 52) meshes with the larger differential spur gears (30 & 41, 31 & 51). The larger differential spur gears are integral with the smaller differential spur gears. To lock the differential, the shift bracket (26 & 45) on the driveshaft (25 & 51) for the sidecar is pushed so that the shift bracket hooks onto the opposing bracket (44) of the housing. Thus any turning of the differential gear casing opposed to the driveshaft (25 & 54) is no longer possible, and the latter must turn at the same speed as the drive wheel (36 & 48) by which the rear wheel is driven.

Zum Ausgleich der verschiedenen Drehzahlen auf Hinterrad und Seitenwagenrad bei Kurvenfahrten ist ein Radantrieb mit Ausgleichgetriebe eingebaut. Das Ausgleichgetriebe ist ein Stirnradgetriebe mit ungleicher Drehmomentverteilung (siehe auch D. »Das Drehmoment«). Es läßt sich sperren, um das Durchgehen eines Rades bei verschlammten Straßen sowie im Gelände zu verhindern.

Das Antriebskegelrad (19, 35 u. 38) treibt das mit dem inneren Ausgleichgetriebegehäuse (10 u. 43) verbundene Tellerrad (18, 34 u. 39) an. Das Antriebsdrehmoment wird im Ausgleichgetriebe über je ein gleiches Vorgelege [zum Hinterrad: Antriebsrad (17, 36 u. 48) und Mitnehmerrad (14, 33 u. 47), zum Seitenwagenrad: Antriebsrad (55) und Mitnehmerrad (46)] auf Hinterrad und Seitenwagenrad

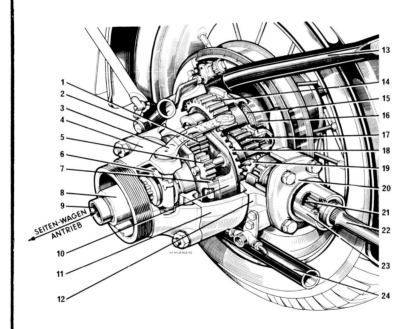

1 Ausgleichstirnrad
2 Befestigungsschraube
3 Sperrhebel
4 Deckel für Hauptanschluß
5 Ausgleichzwischenrad
6 Schaltgabel für Sperre
7 Innerer Deckel
8 Mitnehmerglocke
9 Mitnehmer für Seitenwagenantrieb
10 Inneres Ausgleichgetriebegehäuse
11 Äußerer Gehäusedeckel
12 Äußeres Ausgleichgetriebegehäuse
13 Rahmenrohr
14 Mitnehmerrad
15 Mitnehmer
16 Nabe
17 Antriebsrad
18 Tellerrad
19 Antriebskegelrad
20 Lagerflansch
21 Mitnehmer
22 Gelenkwelle
23 Mitnehmerglocke
24 Rahmenrohr

verteilt. Das Stirnrad (32 u. 50) für Antrieb des Hinterrades greift in die kleineren Ausgleichstirnräder (30 u. 40 sowie 31 u. 49) ein, während das Stirnrad (28 u. 53) für Antrieb des Seitenwagens über die Ausgleichzwischenräder (27 u. 42 sowie 29 u. 52) mit den größeren Ausgleichstirnrädern (30 u. 41 sowie 31 u. 51) in Verbindung steht. Die größeren Ausgleichstirnräder sind mit den kleineren Ausgleichstirnrädern aus einem Stück. Zum Sperren des Ausgleichgetrie-

bes wird die Schaltklaue (26 u. 45) auf der Antriebswelle (25 u. 54) für Seitenwagen derart verschoben, daß die Schaltklaue in die Gegenklaue (44) des Gehäuses eingreift. Hierdurch ist ein Verdrehen des Ausgleichgetriebegehäuses gegenüber der Antriebswelle (25 u. 54) nicht mehr möglich, und diese muß mit der gleichen Drehzahl laufen, wie das Antriebsrad (36 u. 48), über welches das Hinterrad angetrieben wird.

LEHRTAFEL 32 (ERLÄUTERUNG)   BAYERISCHE MOTOREN WERKE AG., MÜNCHEN

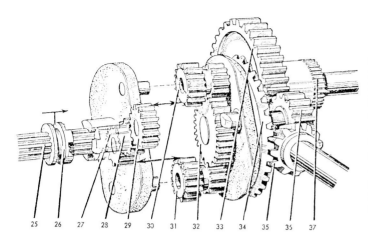

25 Antriebswelle für Seitenwagen
26 Schaltklaue für Ausgleichgetriebesperre
27 Ausgleichzwischenrad
28 Stirnrad für Antrieb des Seitenwagens
29 Ausgleichzwischenrad
30 Ausgleichstirnrad
31 Ausgleichstirnrad
32 Stirnrad für Antrieb des Hinterrades
33 Mitnehmerrad
34 Tellerrad
35 Antriebskegelrad
36 Antriebsrad
37 Mitnehmer

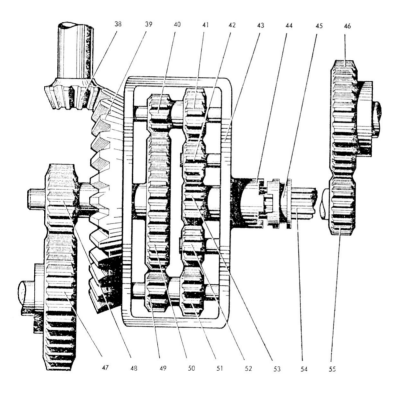

38 Antriebskegelrad
39 Tellerrad
40 Kleines Ausgleichstirnrad
41 Großes Ausgleichstirnrad
42 Ausgleichzwischenrad
43 Ausgleichgetriebegehäuse
44 Gegenklaue für Ausgleichgetriebesperre
45 Schaltklaue für Ausgleichgetriebesperre
46 Mitnehmerrad im Seitenwagenantrieb

47 Mitnehmerrad im Hinterradantrieb
48 Antriebsrad im Hinterradantrieb
49 Kleines Ausgleichstirnrad
50 Stirnrad für Antrieb des Hinterrades
51 Großes Ausgleichstirnrad
52 Ausgleichzwischenrad
53 Stirnrad für Antrieb des Seitenwagens
54 Antriebswelle für Seitenwagen
55 Antriebsrad im Seitenwagenantrieb

1. Differential spur gear
2. Attachment screw
3. Locking lever
4. Cover for main connection
5. Differential intermediate gear
6. Shifting fork for lock
7. Inner cover
8. Carrier housing
9. Carrier for sidecar drive
10. Inner differential gear housing
11. Outer housing cover
12. Outer differential gear housing
13. Frame member
14. Carrier gear
15. Carrier
16. Hub
17. Drive gear
18. Crown gear
19. Drive bevel gear
20. Bearing flange
21. Carrier
22. Driveshaft
23. Carrier housing
24. Frame member
25. Driveshaft for sidecar wheel
26. Shift bracket for differential lock
27. Intermediate differential gear
28. Spur gear for sidecar drive
29. Intermediate differential gear
30. Differential spur gear
31. Differential spur gear
32. Spur gear for rear wheel drive
33. Carrier gear
34. Crown gear
35. Drive bevel gear
36. Drive gear
37. Carrier
38. Drive bevel gear
39. Crown gear
40. Small differential spur gear
41. Large differential spur gear
42. Intermediate differential gear
43. Differential gear unit
44. Opposing bracket for differential lock
45. Shift bracket for differential lock
46. Carrier gear for sidecar drive
47. Carrier gear for rear wheel drive
48. Drive gear for rear wheel
49. Small differential spur gear
50. Spur gear for rear wheel drive
51. Large differential spur gear
52. Intermediate differential gear
53. Spur gear for sidecar drive
54. Sidecar driveshaft
55. Sidecar drive gear

The differential divides the torque and also equalizes the wheel speeds; since the two occur independently of each other, the description will also be offered separately so as to be easier to understand.

A. Turning Speed Equalization:
When driving straight ahead, none of the gears in the differential turns. The differential spur gears thus turn around the drive spur gears without being turned by them, so that the drive spur gears must turn at the same speed as the entire unit.

In a curve to the right, the sidecar wheel turns more slowly than the rear wheel, because the inside curve is shorter than the outside curve. This can theoretically be taken to the point at which the sidecar wheel stops. This situation is dealt with as follows:
The sidecar wheel stands still, that means the carrier gear (46) and the drive wheel (55) as well as the spur gear (28 & 53) for sidecar drive stand still too. When the differential unit (43) linked to the crown gear (18, 34 and 39) turns, then the intermediate differential gears (27 & 42, 29 & 52) turn like planets around the spur gear (28 & 53). The turning speed of the intermediate differential gears is transmitted to the differential spur gears, which in turn drive the spur gear (32 & 50) for the rear wheel drive, so that the latter experiences a second impetus in addition to the speed of the motion straight ahead, and the rear wheel turns correspondingly faster (nc is greater than nb). In a curve to the left, the same process occurs as in a curve to the right.

B. Torque Division:
The center of gravity of the vehicle lies nearer to the rear wheel than to the sidecar wheel (the sidecar is considerably lighter than the motorcycle itself), so that the wheel pressures and the resistance to motion that result are different. If the sidecar wheel were given the same torque as the rear wheel, then the power to the sidecar wheel, which is attached to a larger lever arm than the power of the rear wheel, gives the vehicle a push to the left. The differential thus cannot divide the torque in a ratio of 1 : 1 as in a symmetrical vehicle, but rather in an inverse ratio of the lever arms (the distances of the rear wheel and the sidecar wheel from the center of gravity of the vehicle). On the other hand, the wheel's running too far and too fast influences the performance of the vehicle, so that the exact proportions of the torque must be determined through careful experimentation.

Das Ausgleichgetriebe wirkt drehmomentverteilend, in Kurven außerdem drehzahlausgleichend; da beides unabhängig voneinander geschieht, soll die Beschreibung auch getrennt erfolgen, zumal sie dadurch leichter verständlich wird.

A. Drehzahlausgleich:

Bei Geradeaus-Fahrt dreht sich keines der im Ausgleichgetriebegehäuse befindlichen Räder. Die Ausgleichstirnräder laufen also um die Antriebsstirnräder herum, ohne sich auf diesen abzuwälzen, so daß die Antriebsstirnräder mit derselben Drehzahl umlaufen müssen wie das ganze Gehäuse ($n_b = n_b$).

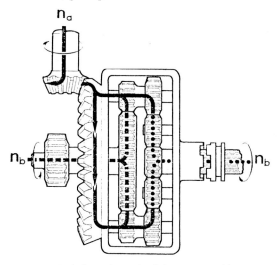

In einer Rechtskurve läuft das Seitenwagenrad langsamer als das Hinterrad, weil die Innenkurve kürzer ist als die Außenkurve. Dies kann theoretisch soweit getrieben werden, daß das Seitenwagenrad stillsteht. Dieser Fall ist im Folgenden zugrunde gelegt worden:

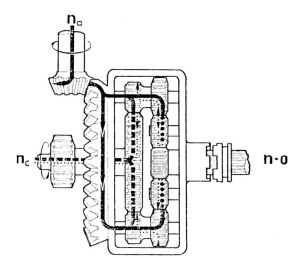

Das Seitenwagenrad steht still, d. h. das Mitnehmerrad (46) und das Antriebsrad (55) sowie das Stirnrad (28 u. 53) für Antrieb des Seitenwagens stehen ebenfalls still. Wenn sich das mit dem Tellerrad (18, 34 u. 39) verbundene Ausgleichgetriebegehäuse (43) dreht, so laufen die Ausgleichzwischenräder (27 u. 42 sowie 29 u. 52) wie Planeten um das Stirnrad (28 u. 53) um. Die Umfangsgeschwindigkeit der Ausgleichzwischenräder wird auf die Ausgleichstirnräder übertragen, die ihrerseits wieder das Stirnrad (32 u. 50) für Antrieb des Hinterrades antreiben, so daß dieses außer der Geschwindigkeit bei Geradeaus-Fahrt einen zusätzlichen Antrieb erfährt und das Hinterrad sich entsprechend schneller dreht (n_c ist größer als n_b). In einer Linkskurve geschieht sinngemäß das gleiche wie in der oben beschriebenen Rechtskurve.

B. Drehmomentverteilung:

Der Schwerpunkt des Gespannes liegt näher am Hinterrad als am Seitenwagenrad (der Seitenwagen ist wesentlich leichter als das eigentliche Kraftrad!), so daß die Raddrücke und die dadurch hervorgerufenen Fahrwiderstände verschieden sind. Würde dem Seitenwagenrad dasselbe Antriebsdrehmoment zugeleitet werden wie dem Hinter-

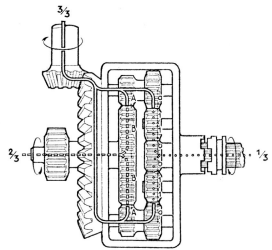

rad, so würde die Kraft am Seitenwagenrad, die an einem größeren Hebelarm angreift als die Kraft am Hinterrad, dem Gespann einen Zug nach links geben. Das Ausgleichgetriebe darf deshalb das Drehmoment nicht wie bei einem symmetrischen Wagen im Verhältnis 1 : 1 aufteilen, sondern im umgekehrten Verhältnis der Hebelarme (der Abstand des Hinterrades bzw. des Seitenwagenrades vom Schwerpunkt des Gespannes). Anderseits beeinflussen Vorspur und Voreilung die Fahreigenschaften des Gespannes, so daß das genaue Verhältnis der Drehmomentverteilung durch sorgfältige Versuche ermittelt werden muß.

Die kleineren Ausgleichstirnräder (40 u. 49) greifen in das große Stirnrad (50) für Antrieb des Hinterrades ein. Die Kraft zum Hinterrad wird also im umgekehrten Verhältnis deren Halbmesser übertragen. Die größeren Ausgleichstirnräder (41 u. 51) greifen in die Ausgleichzwischenräder (42 u. 52) und damit in das Stirnrad (53) für Antrieb des Seitenwagens ein. Da in der Kraftübertragung die Zwischenräder (42 u. 52) wirkungslos bleiben, weil sie nur zur Umkehr der Drehrichtung erforderlich sind, wird das Drehmoment zum Seitenwagenrad im umgekehrten Verhältnis der Halbmesser der größeren Ausgleichstirnräder (41 u. 51) und des Stirnrades (53) für Antrieb des Seitenwagens übertragen. Mit anderen Worten, durch die Kraftübertragung der kleineren Ausgleichstirnräder entsteht am großen Stirnrad für Antrieb des Hinterrades ein großes Drehmoment (etwa $2/3$), während durch die Kraftübertragung der größeren Ausgleichstirnräder nur ein kleines Drehmoment (etwa $1/3$) am kleinen Stirnrad für Antrieb des Seitenwagens entsteht. Dabei ist es gleichgültig, ob sich die Zahnräder im Ausgleichgetriebe gegenseitig verdrehen oder nicht, d. h. ob das Ausgleichgetriebe drehzahlausgleichend wirken muß oder nicht (siehe auch E. »Die Ausgleichwirkung«).

C. Gesperrt:

Wird das Ausgleichgetriebe gesperrt, um das Durchgehen eines Rades zu verhindern, so können sich die Räder im Ausgleichgetriebe gegenseitig nicht mehr verdrehen, so daß das Ausgleichgetriebe die verschiedenen Drehzahlen von Hinterrad und Seitenwagenrad bei Kurvenfahrten nicht mehr ausgleichen kann. Dies bedeutet aber, daß die Lenkfähigkeit des Gespannes beeinträchtigt wird, deshalb ist die Sperre sofort nach Überwinden des Hindernisses wieder zu lösen, so daß die Vorteile des drehzahlausgleichenden Ausgleichgetriebes wieder wirksam werden können.

Die drehmomentverteilende Wirkung des Ausgleichgetriebes wird durch die Sperre nicht beeinträchtigt, da wohl die Welle zum Antrieb des Seitenwagens mit dem Gehäuse gekuppelt wird, nicht aber die Welle zum Antrieb des Hinterrades. Die Kräfte müssen also nach wie vor über die Räder des Ausgleichgetriebes geleitet werden.

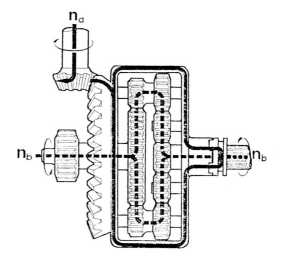

The smaller differential spur gears (40 & 49) mesh with the large spur gear (50) that drives the rear wheel. The power to the rear wheel is thus transmitted in an inverse ratio to its radius. The larger differential spur gears (41 & 51) mesh with the intermediate differential gears (42 & 52) and thus with the spur gear (53) that drives the sidecar wheel. Since the intermediate wheels (42 & 52) remain without effect in power transmission, because they are needed only to change the direction of turning, the torque is transmitted to the sidecar wheel in an inverse ratio of the radius of the larger differential spur gears (41 & 51) and the spur gear (53) that drives the sidecar wheel. In other words, through the power transmission of the smaller differential spur gears, a large torque occurs in the large spur gear that drives the rear wheel (approximately 2/3), while through the power transmission of the larger differential spur gears, only a small torque (approximately 1/3) occurs in the small spur gear that drives the sidecar wheel. Thus it is immaterial whether or not the gears in the differential turn reciprocally, that is, whether or not the differential must equalize the turning speeds (see also E, "The Equalizing Effect").

C. Locked:

When the differential is locked to stop the further turning of one wheel, then the gears in the differential can no longer turn reciprocally, so that the equalization of the different turning speeds of the rear wheel and the sidecar wheel on curves can no longer take place. But this means that the steering potential of the vehicle is influenced, and for that reason the locking must be released again immediately after the hindrance is overcome, so that the advantages of the equalizing differential can become effective again.

The torque-dividing effect of the differential is not influenced by the lock, since the shaft that turns the sidecar wheel is coupled with the unit but not the shaft that turns the rear wheel. Thus the forces must now, as before, be transmitted over the gears of the differential.

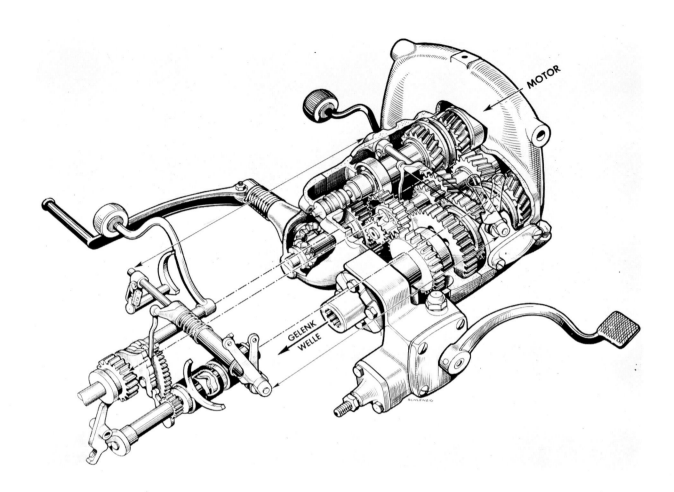

Cutaway drawing of the BMW three-shaft gear drive with starter lever, foot shift and, at lower right, the main cylinder for the hydraulic brakes with foot-pedal activation. On the driveshaft in an extension of the motor's axis, the gear for "road" or for "off-road" is engaged.

Numbers built in the years 1940-1944: 16,545 machines. This vehicle was not exactly cheap. The heavy R 75 motorcycle with everything included cost almost twice as much as a VW personnel car, which offered not only better protection from wind but also much more versatile use.

Technical description of the BMW R 75 (according to description of equipment and instructions for use of June 27, 1941):

Variable Gears:
1. Reverse gear when the off-road shift lever is shifted to "road." The gear ratios are decreased by shifting the off-road shift lever to "off-road." Up to motor number 758,015, at the "off-road" setting only the first three gears were changed, while when 4th gear was engaged, the off-road shift lever came out of the "off-road" position.

Driving farther is then only possible when the off-road shift lever is shifted to the "road" position. The forward gears are engaged with either the hand or foot shift lever. The hand shift lever also serves as a gear indicator. Reverse gear can only be engaged by using the hand shift lever after releasing the shift lock.

The driveshaft connected to the motor drives the main driveshaft via the camshaft, the gears for the forward speeds are constantly meshed with each other in pairs and are always connected to the shaft by a shift bracket. The same thing happens in the gears for road and off-road driving. The two reverse gears are not meshed with each other, but are linked by the intermediate gear when engaged, whereby the turning direction of the main shaft is reversed.

By pushing the foot shift down or letting it up, the shifting roller is moved over the shifting segment, whereby the shifting forks are pushed along curved grooves and the pushing brackets come into contact. The foot shift lever always returns to its initial position. As opposed to that, the hand shift lever always remains in its position at that time. The shifting process in the gearbox is nevertheless the same as with foot shifting. The engaging process is built into the gearbox cover and linked with the second shaft. The gearbox is bolted to the motor to form a single block.

Wheel Drive with Differential and Sidecar Wheel Drive
The wheel drive is attached to the rear part of the frame, the sidecar drive with the flange mount is attached movably to the sidecar frame. The driveshaft is connected to the drive bevel gear by which the crown gear is driven. The crown gear is connected to the differential by which the drive gear and thereby the carrier gear and the rear wheel are driven. Drive of the sidecar wheel takes place via the sprung driveshaft connected to the differential drive, the drive gear and the carrier gear.

The differential is intended to equalize the different turning speeds of the rear wheel and sidecar wheel on curves. The differential is a power-dividing spur gear drive that simultaneously equalizes the unequal resistance caused by the different pressures.

Through the different sizes of the spur gears it is achieved that on a flat, dry pavement correspondingly greater power is transmitted to the rear wheel than to the sidecar wheel.

The differential locks in order to prevent the further turning of one wheels on a muddy road or off the road. For this purpose the shift bracket on the driveshaft is moved so that it contacts the opposite bracket in the differential for equalization. Thus a turning of the differential unit in opposition to the driveshaft is made impossible, and the driveshaft to the sidecar wheel must turn at the same speed as the drive bevel gear by which the rear wheel is driven.

Fork and Front Suspension
The front fork consists of two columns which are screwed through crosspieces (the steering plate and the fork guide) above and below the steering head. Each half of the fork consists of a fixed leading tube (the fork tube), over which the sliding tube (fork endpiece) is pushed.

They are guided by the upper and lower guide boxes, the seal by the pressure-fast sealing ring. The fork tube is braced by a coil spring, its lower end enclosed in the tension piece of the fork endpiece and its upper end in the tension piece of the fork guide, so that the spring can be influenced by both pushing and pulling, thus handling both bumps in the road and recoiling. A double-acting hydraulic damper in each half of the fork dampens the driving vibrations.

The shock absorber consists of a shock-absorber shaft screwed onto the closing screw of the forkand carrying the shock-absorber valve on its lower end, out of the shock-absorber tube attached to the fork endpiece and the guide box, which along with the shock-absorber rod forms the ring-shaped shock-absorber jet. Down below in the shock-absorber tube is the bottom valve, so that oil can be sucked out of the space outside the shock-absorber tube.

Sidecar

The sidecar frame is made of tubes welded in a rectangle. It is attached to the motorcycle by a ball connection on the motor bolt, a hasp bolt on the wheel drive and two adjustable struts. Through the rear cross member the sprung driveshaft and the tube spring, protected from dust, lead to the swing arm, which it attached to the tube by a flange mount.

The sidecar wheel is attached to the swing arm and thus sprung. The sidecar wheel is hung at the rear of the frame on two leaf springs which run forward to a rubber mount. Behind the seat is the luggage space, the cover of which carries the spare wheel. The pack carriers are attached to both sides of the front part.

Foot Shifting

The path of the foot shift lever is limited by the two contact points. Shifting difficulties can be avoided by means of the contact screws as follows: Unscrew the porthole cover and unscrew both contact screws a few turns after loosening the restraining nut with a wrench (Matra No. 477).

Upper Contact: Turn the shafts and engage first gear with the foot shift lever (check porthole). In so doing, push the foot shift lever down until you feel contact and hold it firmly in this position. Screw in the upper contact screw until the foot pedal begins to rise. Turn the contact screw back a quarter turn and fasten it.

Testing: Push the intermediate lever on the gearbox lightly forward and pull the foot pedal strongly upward out of its resting position. A little play must be tangible in the intermediate lever. If the play is too great, the contact screw has been screwed out too far; if no play is tangible, then the contact screw has been screwed in too far. Lower contact: Turn the shafts and engage fourth gear with the foot shift lever (check porthole). In so doing, push the foot shift lever upward until you feel contact and hold it firmly in this position. Screw the lower contact screw in until the foot pedal begins to move downward. Turn the contact screw back a quarter turn and fasten it in this position. Testing: Push the intermediate lever on the gearbox lightly backward and pull the foot shift lever strongly downward out of its resting position. A little play must be tangible in the intermediate lever. If the play is too great, the contact screw has been screwed out too far; if no play is tangible, the contact screw has been screwed in too far.

1. Driveshaft for the sidecar
2. Switch bracket for the differential lock
3. Intermediate differential gear
4. Spur gear for sidecar wheel drive
5. Intermediate differential gear
6. Differential spur gear
7. Differential spur gear
8. Spur gear for rear wheel drive
9. Carrier gear
10. Crown gear
11. Drive bevel gear
12. Drive gear
13. Carrier gear

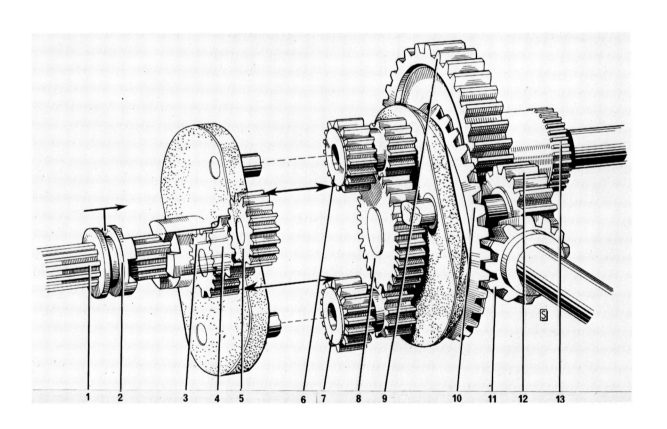

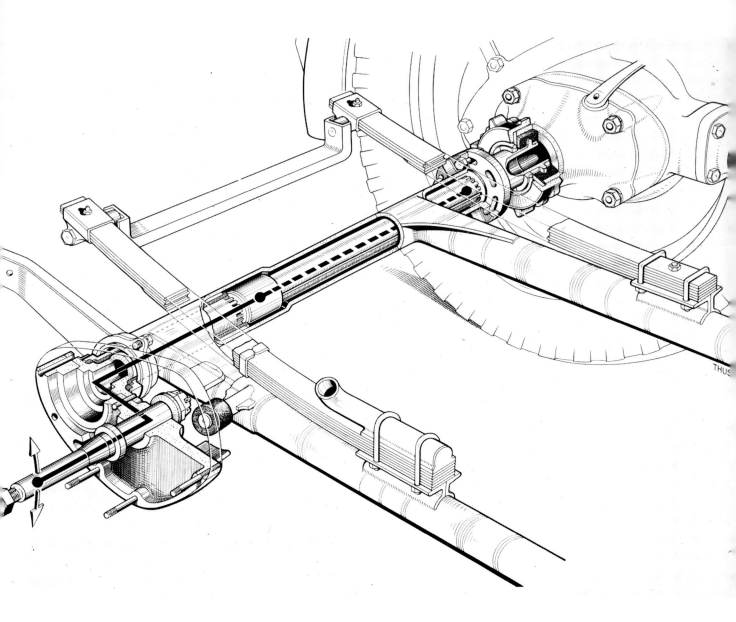

This drawing shows the slit spring tube leading through the sidecar frame, to which the swing arm for the driven sidecar wheel is linked.
Dotted line: spring tube.
Solid black line: Swing arm.

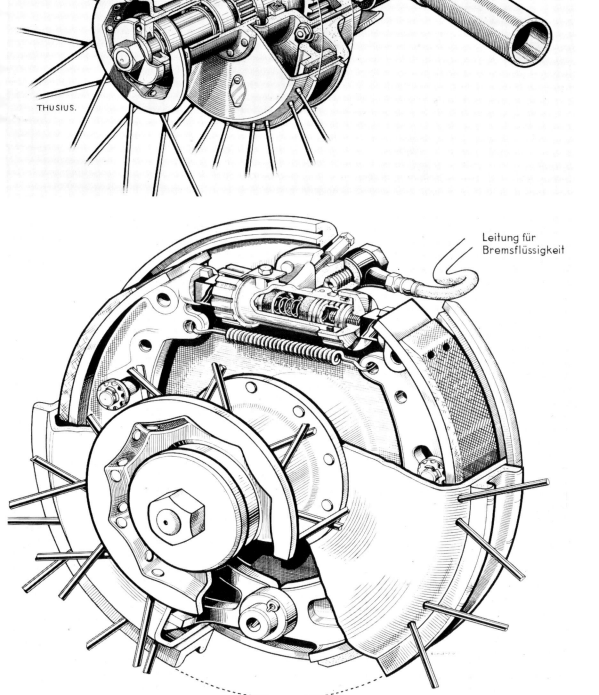

Sidecar drive

Drawing of the brake cylinder in the inboard drum brake of the sidecar wheel of the BMW R 75.

Leitung für Bremsflüssigkeit

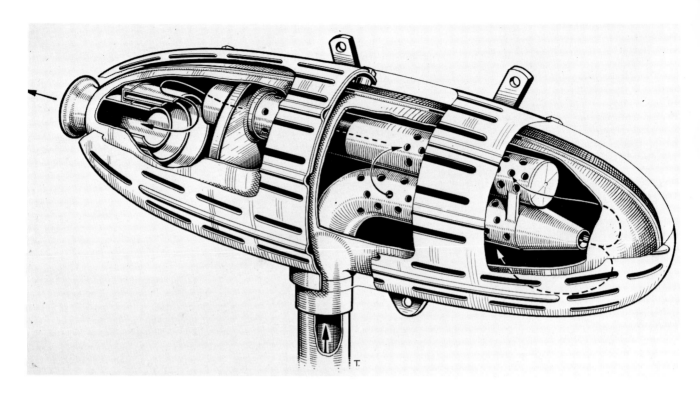

Rear exhaust muffler of a BMW R 75: The exhaust passes through the exhaust pipe into the muffler; from their it is conducted to the rear opening and exits.

Eight main parts form the demountable frame of the BMW R 75. Each part can also be replaced independently, which means not only smaller storage space but also much saving of repair time.

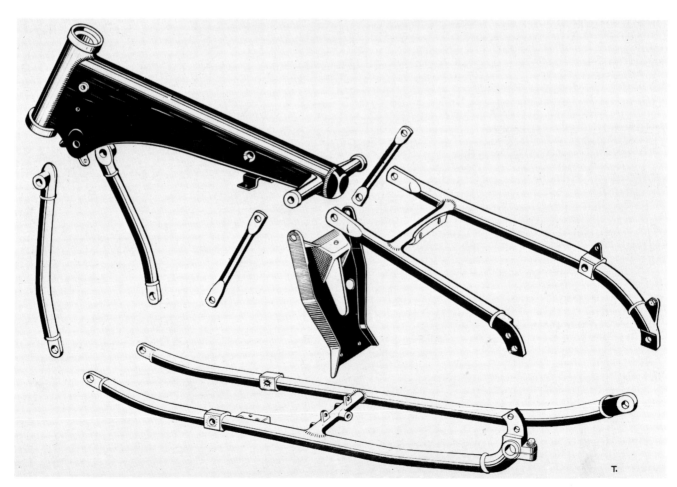

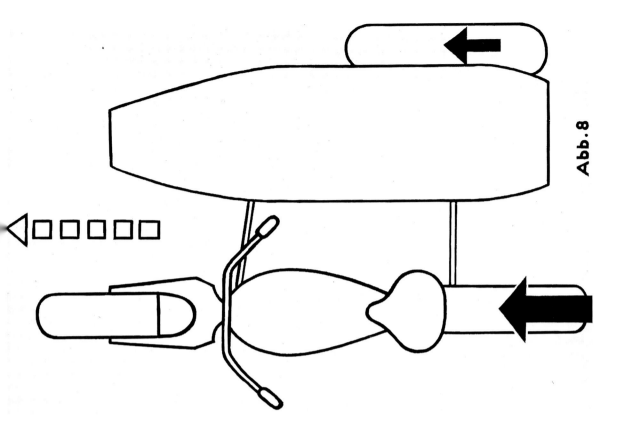

Thanks to the sidecar being driven by lesser torque than the rear wheel, the excess speed of the sidecar and thus the lateral pressure on the steering is restrained. The vehicle runs straight ahead with no trouble.

Abb. 8

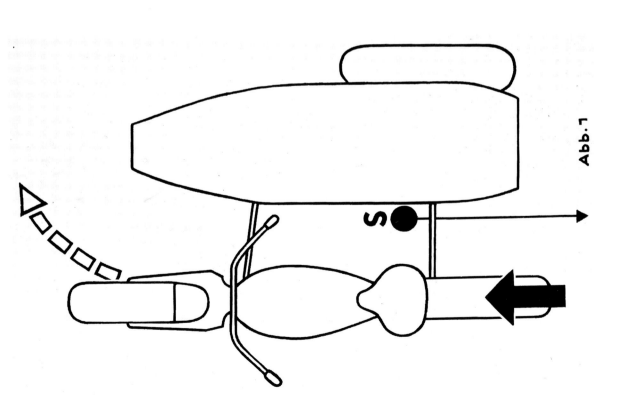

The center of gravity of a normally propelled motorcycle and sidecar lies to the side near the cycle and, through its inertial resistance, produces an opposing force that tends to hold back the sidecar and give the vehicle a slight pull to the right.

Abb. 7

Oil circulation

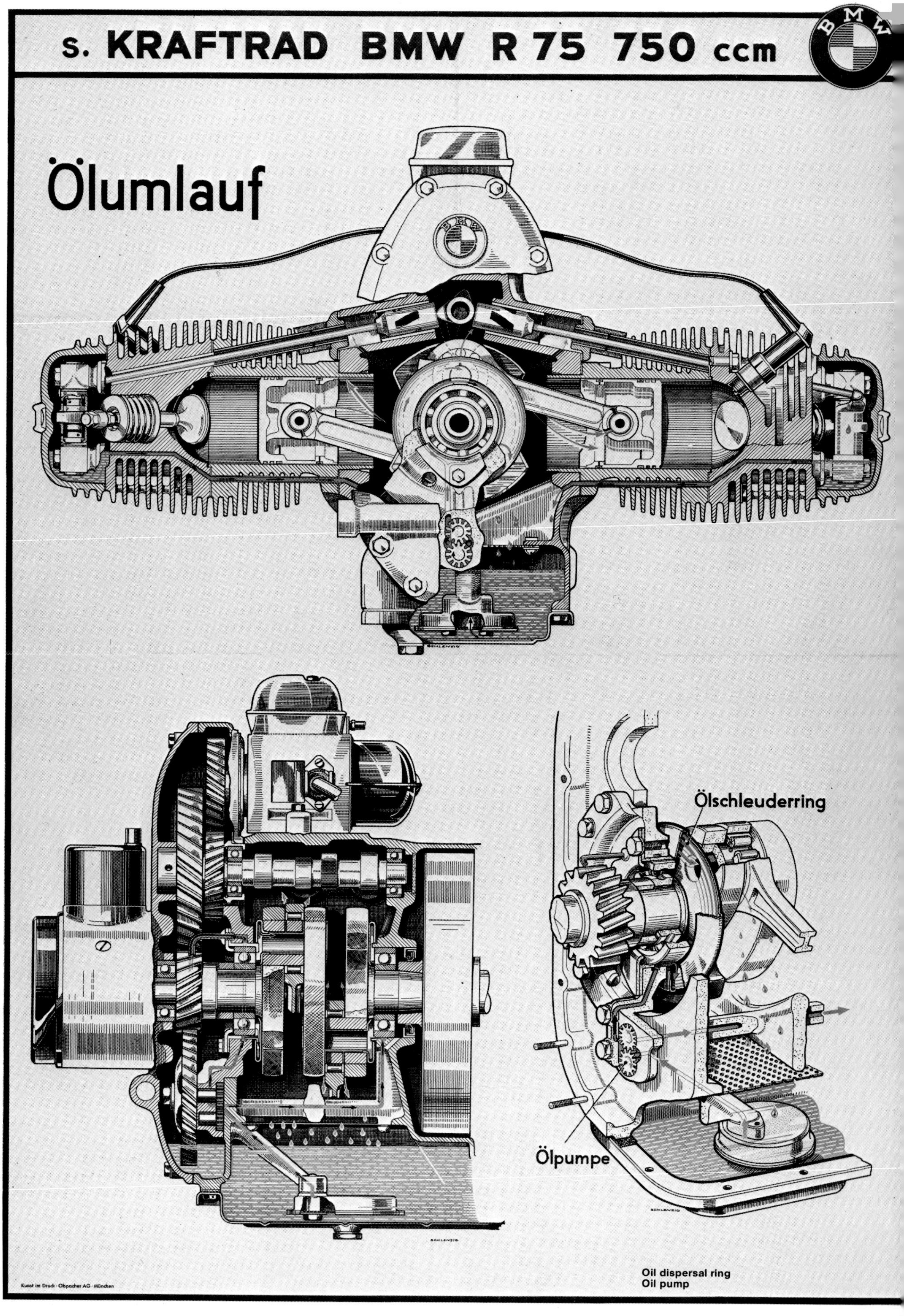

s. KRAFTRAD BMW R 75 750 ccm
BMW
Ölumlauf
Ölschleuderring
Ölpumpe
Kunst im Druck · Obpacher AG · München

Oil dispersal ring
Oil pump

A WORD OF THANKS

I would like to express my hearty thanks for their help to:

Mr. C. Grosser, Overath
Mr. J. Fleischmann, BMW Archives, Munich
Mr. H. Dost, Solingen
Dr. M. Haupt, Federal Archives, Koblenz
All the personnel of the Photographic Library, Imperial War Museum, London
Mr. J. S. Lucas and Mr. P. H. Reed, Imperial War Museum, London

BIBLIOGRAPHY

Bayerische Motorenwerke: Repair manual for BMW R 75 Motorcycle, Munich 1942

Benary, Albert, *Die Berliner 257, Bear Division*, Bad Nauheim 1957

Breithaupt, Hans, *Die Geschichte der 30. Infanterie-Division*, Bad Nauheim 1955

Conze, W., *Die Geschichte der 291. Infanterie-Division*, Bad Nauheim 1953

Dieckhoff, G., *Die 3. Infanterie-Division (mot.)*, Göttingen 1960

Ellis, Chris, *Military Transport of World War II*, London 1970

Hausser, Paul, *Waffen-SS im Einsatz*, Göttingen 1953

Hubatsch, Walter, *61. Infanterie-Division*, Bad Nauheim 1958

Janssen, Gregor, *Das Ministerium Speer*, Berlin 1968.

Meyer, Kurt (Panzer-Meyer), *Grenadiere*, Munich 1965

Mueller-Hillebrand, Burkhart, *Das Heer 1933-1945*, Vol. 1-3, Frankfurt 1954-1969

OKH, *Heeres-Waffen-Amt, Schweres Kraftrad 750 cc mit Seitenwagen (angetrieben) BMW, Baumuster 750/275 (bisherige Bezeichnung R-75), Gerätebeschreibung und Bedienungsanweisung*, Berlin 1943

Reinicke, Adolf, *Die 5. Jäger-Division*, Bad Nauheim 1962

Spaeter, Helmuth, *Die Geschichte des Panzerkorps Grossdeutschland* (privately published)

Speer, Albert, *Erinnerungen*, Berlin 1968

Walther, Herbert, *Die Waffen-SS*, Echzell 1972

Werthen, Wolfgang, *Geschichte der 16. Panzer-Division*, Bad Nauheim 1958

PERIODICALS

Der Adler, 1938-1944
BMW-Werkzeitung, 1941-1943
Der Frontsoldat erzählt, 1953-1960
Das Heer, 1938-1944
Die Wehrmacht, 1938-1944
Wehrwissenschaftliches Rundschau, 1950-1975

PHOTO SOURCES

BMW Archives (28)
Federal Archives, Koblenz (205)
Imperial War Museum, London (9)
J. Piekalkiewizc Archives (11)

STATEMENTS

In the author's possession.

German Military Rifles and Machine Pistols 1871-1945

Hans Dieter Götz

This richly illustrated volume portrays the development of the modern German weapons and their ammunition, and includes many rare and experimental types. Among the weapons covered in this book are the Werder rifle, Mauser rifles, the various M/71 rifles and ammunition, the 88 cartridge, the infantry rifle 88, ther 98 rifles, the 41 & 43 rifles, the Fallschirmjäger rifle, ERMA & Walther assault rifles, the STEN copy and many more. Value guide included.
Size: 8 1/2" x 11"   248 pp.
Many documents, line drawings, over 200 photos
ISBN: 0-88740-264-X   hard cover $35.00

The German Infantry Handbook 1939-1945

Alex Buchner

Many millions of soldiers belonged to the more than 500 German infantry divisions between 1939-1945. But was an infantry division like at that time? What sort of form did it take, and how did the structures, the organization, and the equipment of an infantry division look during World War II?

Alex Buchner sets out to give answers to these questions, presenting the first inclusive picture of the German infantry. It illuminates the inner connections and describes everything that made up an infantry division at that time – formation, strength, armament, equipment, rank insignias, rifle groups, rifle columns, the company, light infantry weapons, reconnaissance units, Panzerjäger units, the artillery regiment, engineer battalions, veterinary services, information units, support services, medical services, new infantry weapons, divisional casualties and much more. This is an important work of war history, unfolding a hitherto unseen portrait in text and pictures.

For historians and for those interested in the history of men at war, this volume will become one of the classics in the literature on the German Wehrmacht.

Size: 7" x 10"   over 225 photos   288pp.
ISBN: 0-88740-284-4   hard cover $29.95

First English Edition
Paul Carell's

OPERATION BARBAROSSA IN PHOTOGRAPHS
The War in Russia as Photographed by the Soldiers

These photos are irreplaceable witnesses of the war in the east, and impressively and inclusively illustrate what Carell described so grippingly in words in his superb Russian Front studies, *Hitler Moves East*, and *Scorched Earth*.
   As in his textual volumes, Carell not only utilizes the German sources in this photo volume, but also those of the one time enemy: the photo archives of the Russians. With the support of the Soviet embassy and the Novosti Agency, more than four hundred documentary photos could be obtained from the Soviet Union and evaluated. They show what no German cameraman saw: the other side of the war. This volume of photographs is a genuine Carell work — objective, factual, gripping.
Size: 7" x 10"   458 pp.   Over 570 b/w & color photographs, charts, maps, insignia and divisionsal listings
ISBN: 0-88740-280-1   hard cover $44.95

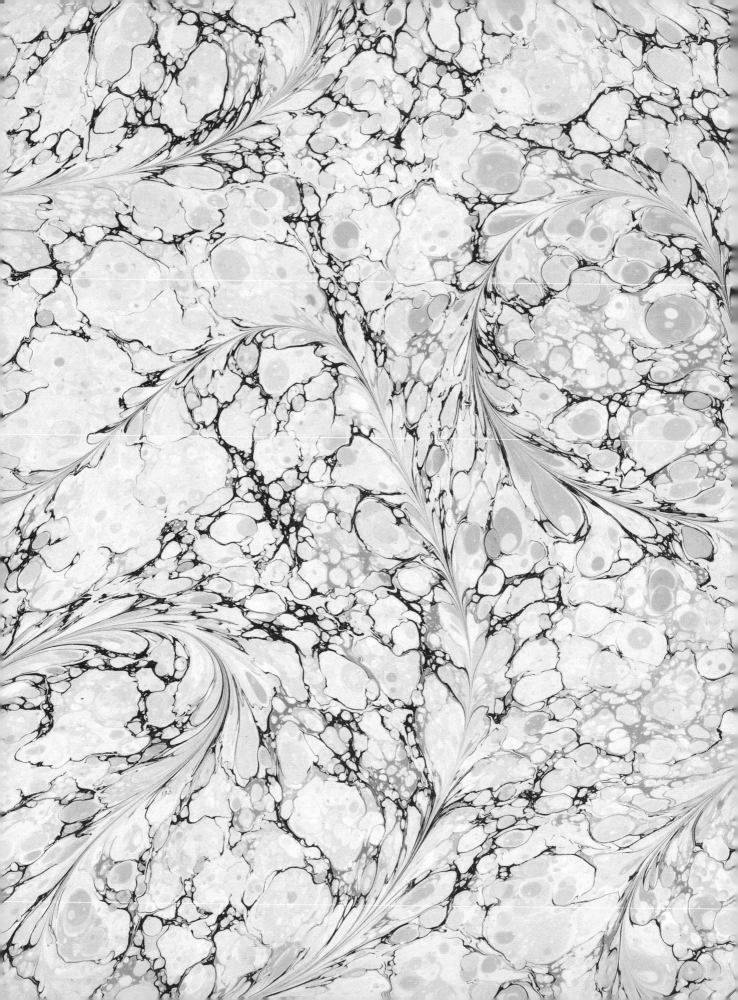